Engine Combustion:
Pressure Measurement and Analysis

David R. Rogers

Warrendale, Pennsylvania
USA

All rights reserved. No part of this publication may be reproduced, stored in a retrieval system, or transmitted, in any form or by any means, electronic, mechanical, photocopying, recording, or otherwise, without the prior written permission of SAE.

For permission and licensing requests contact:

> SAE International Permissions
> 400 Commonwealth Drive
> Warrendale, PA 15096-0001-USA
> Email: permissions@sae.org
> Fax: 724-776-9765

Library of Congress Cataloging-in-Publication Data

Rogers, David (David R.)
  Engine combustion : pressure measurement and analysis / David R. Rogers.
    p. cm.
  Includes bibliographical references and index.
  ISBN 978-0-7680-1963-6
 1. Automobiles—Motors—Cylinders. 2. Automobiles—Motors—Combustion. 3. Pressure—Measurement. I. Title.
  TL214.C93R64 2010
  629.25—dc22   2010021607

> SAE International
> 400 Commonwealth Drive
> Warrendale, PA 15096-0001 USA
> Tel: 877-606-7323 (inside USA and Canada)
> Tel: 724-776-4970 (outside USA and Canada)
> Fax: 724-776-0970
> Email: CustomerService@sae.org
> Website: http://books.sae.org

Copyright © 2010 SAE International
ISBN 978-0-7680-1963-6
DOI 10.4271/R-388
SAE Order No. R-388

Printed in USA

Contents

Chapter 1
A Brief History of Engine Indicators..1

Chapter 2
The Measurement Chain: Encoders..21

Chapter 3
The Measurement Chain: Combustion Pressure Transducers 35

Chapter 4
The Measurement Chain: Additional and Alternative Transducers 91

Chapter 5
The Measurement Chain: Measurement Hardware 111

Chapter 6
The Measurement Chain: Measurement System Software....................145

Chapter 7
Applications..175

Chapter 8
Abnormal Combustion: Measurement and Evaluation 217

Chapter 9
Successful Measurements..255

Chapter 10
Specification and Integration into the Test Environment......................287

Bibliography 303

References 305

Index 309

About the Author 322

Introduction

Indicator, or combustion pressure measurements, used to be the preserve of "experts" only. But, due to the general decrease in the cost of the measurement technology, and the increasing complexity of the task of developing an internal combustion engine—including optimising all the available parameters—combustion pressure measurement is no longer an exclusive task. Today, it is much more likely that test cell operators, technicians, and engineers will be required to have a clear understanding of the measurement procedure and the equipment involved.

The idea behind this book is to provide the inexperienced technical person, trying for perhaps the first time to understand combustion pressure measurements, a source of basic information and guidance on what equipment is available and how to use it in various common applications. The book has been developed as a handbook, to be used as an initial source of information and then, perhaps, to be consulted regularly as a reminder during daily work, or to point the reader to other good sources of information on the subject. This book is not offered as a detailed thesis on the subject of engine combustion and thermodynamics; there are many excellent sources of information available to supplement the information in this book, that will enable readers to develop their knowledge further on the topic of combustion and how to measure, understand, and optimise it.

There is much tribal knowledge in a specialized environment, and best practice, gained from experience, is often shared among users and operators of combustion measurement equipment. This book is an attempt to bring together some of this knowledge and combine it with suggested pathways to further, more detailed information. This should enable readers to go from understanding first principles, to using equipment, gaining experience, and finally to being able to confidently purchase equipment of the correct price and performance for their needs.

The important points to consider with respect to combustion measurements are that this equipment is relatively expensive and sensitive, so that a correct understanding is a prerequisite for an efficient measurement process. It is important always to bear in mind that the only thing produced by combustion measurement is data. If this data is in any way inaccurate, considerable time and money can be wasted. Therefore, familiarity with the measurement system and its efficient operation is essential. In addition, combustion measurements typically create large data files—and while the cost of storing data is relatively low, the time needed to reduce or mine this data, to get the important information from it, can be considerable and should not be overlooked. Data can, of course, be reduced, but it is important to understand the compromises in each stage of reduction, and, if possible, to maintain reversibility and traceability so that if it should be necessary to go backward, post-reduction, deeper into the detail of the raw data, that can still possible.

I would like to thank the companies who have supported me in this project by providing and sharing details of their equipment, systems, and technologies for combustion pressure measurement. I am very grateful to them for this. In particular, I would like to thank Sandra Gildemeister of AVL List GmbH for her ongoing assistance throughout the project.

David R Rogers, 2010

Companies who have provided support and information:

AVL List GmbH (www.avl.com)

Kistler Instruments (www.kistler.com)

d2t (www.d2t.com)

FEV (www.fev.com)

Dewetron (www.dewetron.com)

Lehmann and Michels GmbH (www.lemag.com)

Optrand Inc. (www.optrand.com)

Delphi (www.delphi.com)

Polytec GmbH (www.polytec.com)

Wolff Control Corp (www.wolffcontrols.com)

Jodon Inc. (www.jodon.com)

Chapter 1
A Brief History of Engine Indicators

1.1 Early Engine Indicators

1.1.1 The First Indicators

The process of measuring the pressure inside the cylinder of a reciprocating piston engine dates back to the dawn of the reciprocating engine itself. During the early development of the steam engine by James Watt and others, understanding the in-cylinder process of energy release was fundamental to optimising the performance of the engine as a complete machine. The engine cycle operation can be represented by a diagram that plots instantaneous cylinder pressure against cylinder volume, known as an indicator diagram. Generally the process of taking such measurements uses a device known as an engine indicator, which generates the diagram while the engine is running. Information from these measurements is an essential element in the goal of improving efficiencies and optimising in-cylinder motion and expansion of the working fluids.

In the early days, mechanical devices were used to gather in-cylinder information. These devices, known as indicators, were installed on the engine and were subjected to the working pressure in the cylinder; in addition, they were connected to the engine crosshead or crankshaft. From these two fundamental measurement inputs (pressure and volume, volume derived from the crankshaft/crosshead position) the characteristic loop of the pressure/work cycle could be recorded for analysis.

The indicators used by the steam engine pioneers were very primitive. The first steam engines had no crankshaft. They were generally beam-type engines, and the indicator was attached to the crosshead, which had longitudinal motion. The general arrangement is shown in Figure 1.1. An early improvement was the "moving tablet" indicator. This was a simple design and illustrates the measurement principle with great clarity as shown in Figure 1.2.

The cylinder pressure was applied via a valve to an indicator cylinder and to the moving piston, displacing it accordingly and moving the pencil in proportion to the working pressure. The cord in Figure 1.2 was connected to the moving crosshead or rotating crankshaft, and thus as motion occurred, the tablet would move linearly. The interacting movements of the piston/pencil and the tablet when the engine was running produced a closed-loop diagram that would be drawn on the tablet with each working cycle. This diagram constituted the basic recorded data. Once the measurement was completed, the tablet was removed. With the application of appropriate scaling factors for pressure and volume, the area represented by the closed-loop diagram could be established. Note that finding this loop area satisfies the basic requirement of the measurement technique: to understand the amount of energy released or work done in the cylinder. In fundamental terms, pressure is measured as a function of the volume, and because the area of the recorded loop represents pressure multiplied by volume, the area can also be expressed in units of work. Thus the area of the enclosed loop represents the work done in the cylinder during that cycle. If the engine speed is known, power can be determined and this value can be compared with the actual power output from the engine. Thus, the indicator measurement can help the engineer identify the efficiency of the engine and can assist in further development to minimise losses. This still holds true for today's engineers making indicator measurements. Although the equipment is much more sophisticated, the same basic principles apply and the development goals are the same:

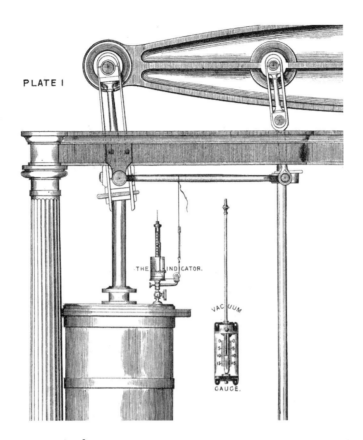

Figure 1.1 Indicator attached to the crosshead of a beam engine.
(Source: John Walter, Archiving Industry.)

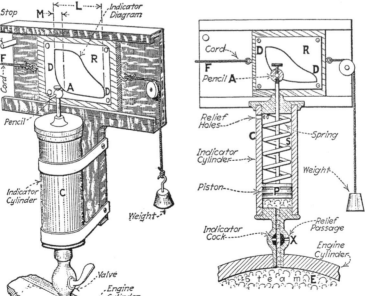

Figure 1.2 Moving tablet type of indicator, circa 1790s.
(Source: John Walter, Archiving Industry.)

to understand the energy release process inside the cylinder with the goal of optimising efficiency.

1.1.2 Early Developments in Indicator Technology

As steam engines evolved, more sophisticated mechanical devices were developed and utilised, ones that could cope with the requirements for increasingly higher engine speeds and pressures. During this time the drum-type indicator was developed, and this basic design has served to execute the measuring task, using a simple mechanical technique, for nearly 200 years. Fundamentally, the same basic design as originally proposed is still in use today, which demonstrates the excellence of the engineering and design practice applied at that time. Figure 1.3 shows the basic construction of the drum-type indicator. Replacing the moving tablet with a rotating drum, the drum-type indicator has the advantage of reduced friction, as well as being more compact.

Cylinder pressure is applied via the tapered plug to the piston in the cylinder of the indicator. As the cylinder pressure increases, force is applied to the piston, which moves against the opposing force provided by the spring surrounding the piston rod. This movement is applied via a delicate linkage to the pencil, so that the amount of pencil deflection is proportional to the cylinder pressure. The rotating drum is spring-loaded in a rest position. The drum base is fastened to a cord that wraps around it. This cord exits the device via a guide pulley and is ultimately attached to the crankshaft at a suitable point. Once the cord is attached and tensioned correctly, rotational movement of the crank will apply a cyclic tensioning of the cord, which rotates the drum, and then releases it. The drum position thus has a direct relationship with the crank position and hence the piston position and cylinder volume. With engine operation, and the proper setup, this allows the two measured parameters of cylinder pressure and crank position (i.e., volume) to be drawn and plotted against each other at the drum surface. The drum is normally loaded with a sheet of paper, held in place with small steel fingers. After a measurement, the recorded curve can be removed from the instrument for analysis and further calculations. A typical indicator diagram from such a device is shown in Figure 1.4.

The spring in the measuring cylinder (opposing the force from the cylinder pressure) is interchangeable, allowing the unit to be used on engines of varying applications (depending

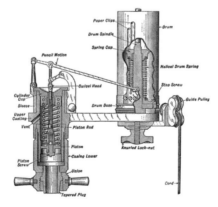

Figure 1.3 A drum-type steam engine indicator.
(Source: www.oldengine.com)

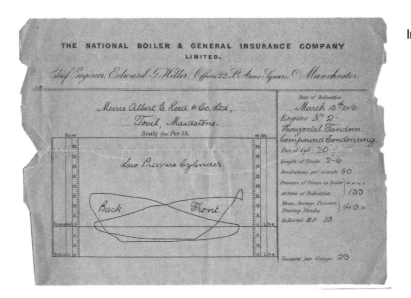

Figure 1.4 Indicator diagram produced by steam engine drum-type indicator.
(Source: John Walter, Archiving Industry.)

on the peak pressure of the engine). Numerous indicator designs were developed to overcome inherent problems associated with measurement, such as pressure oscillations and adaptations to engines with different degrees of crank throw. Although these simple mechanical devices were developed for steam engine applications, as internal combustion engines became more common in the marketplace, the same technology was applied to these engines to assist the engineer in understanding the in-cylinder energy transfer process. A notable development was the Richards indicator, which was developed specifically to accompany the Porter-Allen engine exhibited at the Great Exhibition in 1862 and was a major advance at the time (Figure 1.5).

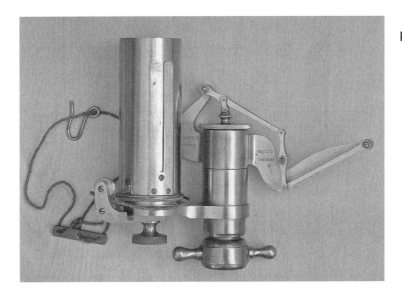

Figure 1.5 The Richards indicator.
(Source: John Walter, Archiving Industry.)

1.1.3 Further Developments in Indicator Technology

During the late 1800s, the internal combustion engine made a heavy impact on the design of the engine indicator. The main change was an increase in the peak cylinder pressures to be measured, in conjunction with the higher number of cylinders in the engine. Smaller-diameter measuring pistons helped handle the new pressure requirements, and special continuous recording designs were developed that could monitor engine performance over extended periods.

Optical indicators were the next evolutionary step in the quest for a more accurate method of producing an indicator diagram. The first generation of optical systems relied on the principle of a pressure-sensing diaphragm, against which the cylinder pressure would be applied via a stopcock. The cyclic rise and fall in pressure would deflect the diaphragm surface, and this movement would be applied to a small mirror. In the perpendicular axis, lateral movement was effected by a mechanism linked ultimately to the crankshaft. This caused biaxial movement of the mirror in relation to cylinder pressure and crank movement, such that when a pinpoint beam of light was directed at the mirror, the reflected, moving beam would produce the equivalent of an indicator diagram that could be displayed on a screen or recorded on a photographic plate. An early optical indicator designed by Charles Bedell is shown in Figure 1.6.

Even though the optical system is capable of responding to high-speed pressure variations, it had the inherent problem of being difficult to calibrate. In addition, projecting the image onto a flat surface affected the ultimate accuracy of such optical systems.

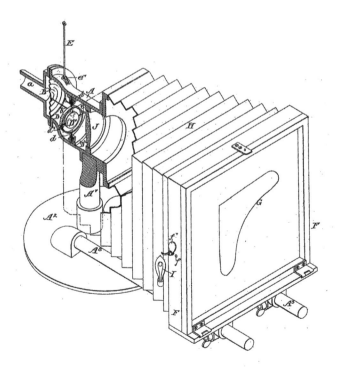

Figure 1.6 A diaphragm-type optical indicator designed by Charles Bedell.
(Source: John Walter, Archiving Industry.)

Many optical systems were proposed between 1900 and 1930, with various technical alterations to overcome specific limitations or accuracy issues. Of note is the Midgely optical, invented by Thomas Midgely, Jr. Midgely was involved in the development of high-speed internal combustion engines and was particularly interested in understanding the phenomenon of abnormal combustion. The available indicator technology at that time was not suitable for his research purpose, so he developed a system of his own, as shown in Figure 1.7.

In his design, Midgely incorporated two mirrors. One was placed to show the changes in pressure by moving the trace vertically. The other was used to display the length of the cycle by deflecting the trace laterally. The vertical movement was effected by a tilting mirror, controlled by a rod attached to the spring-loaded piston that was subjected to the force from the cylinder pressure. The horizontal movement was implemented by a multi-faceted vertical mirror mounted within the wooden body of the instrument. This was positioned in a quadrant that supported the curved viewing screen.

The vertical mirror was driven by a motor, controlled by a drive mechanism, and carried on a separate baseplate so that it rotated in phase with the engine crankshaft. Typically, the motor rotated at one-eighth of the engine speed; each of the eight facets of the mirror, therefore, recorded one complete revolution of the engine. Additional features included the ability to alter the rotation of the mirror in relation to the engine crankshaft, to obtain a photographic record of a single engine cycle. Also, the instrument could be made a pressure × volume recorder simply by allowing the faceted mirror to oscillate instead of rotate.

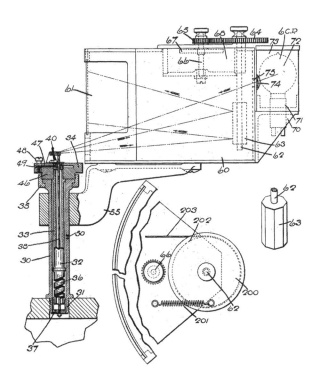

Figure 1.7 Midgely optical indicator.

(Source: John Walter, Archiving Industry.)

The Midgely indicator was a very sophisticated instrument in its day; several patents were filed in relation to this instrument and its application by the employees of General Motors who used it in their engine research and development projects. By the end of the 1930s, however, optical instruments were becoming less favoured, The combination of moving mirrors and mechanical drive mechanisms introduced accuracy problems due to unwanted friction and inertia. The development of the engine indicator was moving toward electrical instrumentation due to the technological advances in electrical and instrumentation engineering occurring at that time.

The next major step forward was the introduction of the spark-trace indicator. This unit is capable of producing large, accurate indicator diagrams over a wide range of engine speeds, but it should be noted that the diagram produced does not represent a single cycle; it is a composite diagram collated from information gathered over a number of successive engine cycles. The system unit consists of a recorder unit and a pickup installed in the engine. It utilises a pressure balance principle to measure the cylinder pressure, which is then recorded on paper via an electrical arc that burns the trace onto the paper surface. A basic system layout for three engine cylinders is shown in Figure 1.8, and a detailed drawing of the actual recorder unit is shown in Figure 1.9.

The drum of the recording unit is driven at engine speed and synchronised with a known crank position. The pressure cylinder piston acts upon the lever, which moves the spark pointer axially across the drum surface. The displacement of the piston occurs as a result of nitrogen gas pressure applied from an external source via a pressure regulator control. Force from the piston movement is opposed by calibrated springs that dictate the scaling factor of the system. The pickup assembly is mounted at the engine cylinder and connected

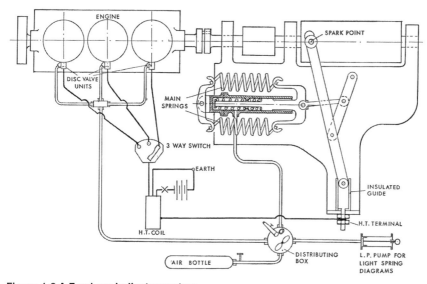

Figure 1.8 A Farnboro indicator system.
(Source: Greene and Lucas. *The Testing of Internal Combustion Engines*. 1969.)

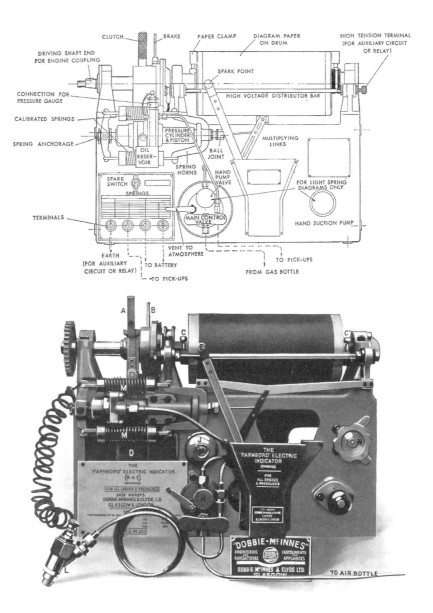

Figure 1.9 Farnboro recorder unit.
(Source: Greene and Lucas. *The Testing of Internal Combustion Engines*. 1969.)

to the combustion chamber pressure via an appropriate tapping. In simple terms, it forms a pressure-controlled contact breaker. It contains a disc mounted on an electrically insulated spindle. The disc, which is exposed to combustion chamber pressure on one side and nitrogen gas pressure on the other, is free to move between two seated positions. The lower position of the seat forms an electrical ground connection. In operation, the disc is held in contact with the lower seat (by the nitrogen gas pressure) until the combustion chamber pressure exceeds the gas pressure. The disc then lifts, breaks contact with the lower seat, and remains in position against the upper seat until the combustion chamber pressure falls again, at which point the disc again contacts the lower seat. Referring to Figure 1.10, it can be seen that a high-tension electrical spark generation circuit is employed to mark the recorder paper wrapped around the drum. The pickup forms part of this electrical circuit (low-tension side) and thus, when the disc moves from the lower seat, the connected primary circuit is broken and a high-tension spark is generated at the pointer, which finds a path to ground via the drum, burning a small hole in the paper as a result. Note that movement of the disk causes the circuit to be interrupted twice per cycle, once when the pressure rises above the applied nitrogen pressure and once when it falls below. Also note that the pressures at the pickup disc and at the pressure cylinder at the recording unit are the same (they are connected together).

Figure 1.10 Farnboro indicator pickup showing disc valve arrangement.
(Source: Greene and Lucas. *The Testing of Internal Combustion Engines*. 1969.)

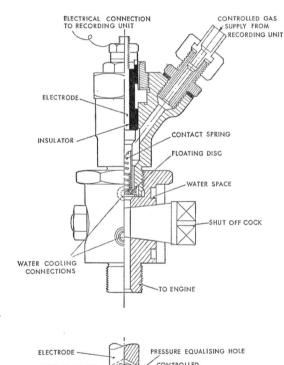

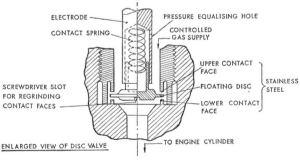

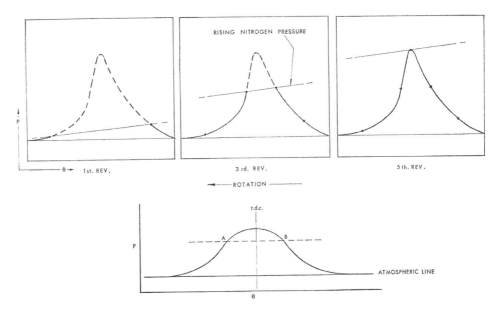

Figure 1.11 Formation of a cylinder pressure diagram using the Farnboro indicator.
(Source: Greene and Lucas. *The Testing of Internal Combustion Engines*. 1969.)

The measurement procedure is as follows, with the engine running and the drum rotating. The pressure control valve is opened gradually so that the nitrogen pressure rises and the spark pointer begins to traverse slowly across the surface of the drum. With every alternate revolution (assuming a four-stroke engine), two sparks will pass from the pointer to ground via the drum, perforating the paper in the process. The first spark is on the compression pressure side, the second on the expansion side of the working pressure cycle of the engine. As the nitrogen pressure increases, the spark pointer moves across the paper, gradually marking out the diagram from the successive pairs of sparks from each cycle. The diagram is complete once the nitrogen pressure exceeds the cylinder pressure. Figure 1.11 illustrates how the diagram progresses during a measurement. After the measurement is completed the paper can be removed from the drum for analysis, and the abscissa can be converted to volume for cylinder work calculations.

Improvements were made to the system over the years, particularly with respect to the pickup, where the original disc valve arrangement could suffer leakage. In later, modified versions, this was replaced by a diaphragm used in conjunction with mains electric high-voltage spark generators. This improved the quality of the final diagram, particularly at high speed, and allowed accurate calibration of the equipment.

An interesting point to note is that the spark-trace type of indicator could be made suitable for "low-pressure" measurements by installing appropriate "light" springs and by replacing the nitrogen gas source with a vacuum pump. This allowed the system to be used to study engine operation during the gas exchange cycle, that is, engine induction and exhaust strokes.

1.1.4 Analysis of Indicator Diagrams

Once the indicator diagram was generated, an analysis was required in a post-processing type of operation. A simple pressure versus crank angle diagram could provide some information via direct analysis of the pressure curve, indicating the rate of pressure rise and maximum cylinder pressures. The most useful information from an indicator diagram could be derived from the cylinder pressure versus volume loop-shaped curve. As mentioned previously, the area enclosed by the loop is a measure of the work done by the expanding gases on the piston. Hence, from this information, engine efficiency and losses can be determined.

A pressure versus crank angle diagram can be converted to a pressure versus volume diagram via a simple tabular method that converts the crank angle position to a piston position, and from this the instantaneous volume can be derived. This can then be plotted as the abscissa in conjunction with the pressure values. This process is tedious to carry out manually, and mechanical calculators and converters were invented to simplify it. Alternatively, atmospheric lines could be drawn on the diagram and from this the mean working pressure of the cylinder could be obtained in any of several ways. The simplest was

Figure 1.12 A typical planimeter.
(Source: John Walter, Archiving Industry.)

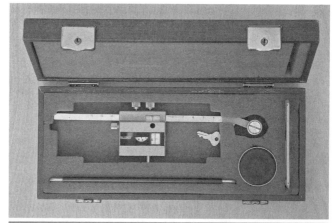

to divide the diagram into narrow vertical sections, total the heights of the sections, and then divide the result by the number of sections. This gave a good approximation, and was made easier by manufacturers who provided grids on the drum paper to facilitate accurate division into sections. The most popular method of diagram analysis involved the use of a geometric tool known as a planimeter, which is shown in Figure 1.12.

This simple device is used to measure the area of two-dimensional shapes and is commonly used in mathematical and cartographical applications. It consists of a linkage mechanism with a pointer that follows the boundary of the target shape—in this case the pressure-volume indicator diagram. The other end of the linkage is in a fixed position (for the case of a polar planimeter). The planimeter uses a measuring wheel that rolls and records in one plane, and skids in the opposite plane. The area of the shape is proportional to the number of turns of the wheel in its course along the perimeter of the diagram. The area of the pressure-volume diagram can be used to determine net work done in the high-pressure part of the cycle, for which the integrated area under the curve represents pressure with respect to changes in volume during compression and expansion phases. The resulting data can be compared with crank-derived measurements (speed and torque) to compare indicated and brake power, and thus identify friction losses.

1.2 The Electronic Age
1.2.1 Oscilloscope Recording

The increasing presence of electronics and electronic instruments in the laboratory had an impact on the engine indicator. Electronic devices became commercially available, most of them based on a cathode ray oscilloscope with additional hardware for measuring crank displacement and cylinder pressure. Unlike the mechanical system, electronic devices allowed the diagram to be displayed on a screen rather than permanently drawn. Consequently, a photographic method of diagram capture was often used for data capture and storage. Systems were available from several manufacturers, but all consisted of the following main components:

- A cathode ray oscilloscope (CRO) to display the image
- A pressure transducer installed in the cylinder complete with appropriate signal conditioning
- A time-sweep or angle-recording mechanism
- If necessary, an optical data capture device

Typically, a standard lab-type CRO (see Figure 1.13) was used for display of the indicator diagram. Of course, this unit would have to be capable of an appropriate data acquisition rate at all required engine speeds (i.e., an appropriate time base sampling). In addition, the display would have to be large enough and bright enough for the application (to be viewable and be photographed). Most important was achieving correlation of data acquisition in the time domain with reference to the angular position of the engine. For this purpose, the oscilloscope was externally triggered via a marker disc and sensor mounted on the engine (most oscilloscopes have an external trigger facility). The sensor would produce a suitable pulse, once per revolution, to trigger the oscilloscope acquisition. Such engine-controlled synchronisation ensures that a steady

Figure 1.13 Lab oscilloscope–based indicator system.
(Source: AVL.)

diagram can be displayed on the screen at any engine speed. For angular correlation, the engine marker disc could be equipped with degree marks at defined intervals. This would provide a signal, from an appropriately fitted engine sensor, of crank degrees at suitable intervals, and often this signal was recorded synchronously with the pressure signal via a two-channel oscilloscope. The resulting diagram would show cylinder pressure in the time domain along with a second channel of crank degree marks. These recorded data could be used to generate new diagrams plotting cylinder pressure versus crank angle and cylinder pressure versus volume. Often, the crank degree trace would incorporate a reference signal showing the top dead centre (TDC) position of the engine.

Cylinder pressure measurement could be done with the installation of an appropriate transducer with measuring face exposed to the internal combustion chamber. Various transducer technologies could be utilised for this task:

- Inductive transducers. Pressure from the combustion chamber applied to a diaphragm causes deflection of the diaphragm. This could be measured via a simple coil and magnet arrangement. The raw signal would be proportional to the velocity of the diaphragm movement and hence to the rate of pressure change—a useful signal for understanding combustion phenomena, as it shows the rapidly changing pressures in certain parts of the engine cycle. However, to produce a pressure diagram, this raw signal must be integrated through signal conditioning.
- Piezoelectric. A quartz crystal and diaphragm (measuring face) can be used to produce an electrical charge signal proportional to diaphragm deflection (and hence cylinder pressure). This technology has now been nearly universally adopted for in-cylinder pressure measurements.
- Variable capacitance. Deflection of the measuring face/diaphragm varies the capacitance of a condenser. This forms part of an oscillator circuit that varies in frequency according to the deflection (i.e., cylinder pressure). The resulting signal is

conditioned and gives a DC voltage in relation to cylinder pressure that can then be applied directly to the oscilloscope input.

- Strain gauge. A change in resistance due to strain can be utilised to give an electrical signal proportional to the deflection of the measuring diaphragm. The strain gauge forms part of a bridge circuit that can incorporate temperature compensation. The raw signal can be amplified and then applied directly to the oscilloscope as a voltage signal representing cylinder pressure.

It is clear that there are numerous sensor technologies available to measure the pressure in the cylinder, each with its advantages and disadvantages. Most require some external signal conditioning, but the fundamental objective for these transducers is to access the cylinder pressure in a nonintrusive way and to produce an electrical signal as a function of the average in-cylinder pressure that is of sufficient amplitude and quality for a successful evaluation.

1.2.2 Digital Systems

Developments in digital electronics and processing technology facilitated the production of electronic measurement equipment specifically designed for the task of engine combustion chamber pressure measurement and analysis. High-speed analogue-to-digital converters allowed storage of the data in electronic form and, more important, allowed processing of the data via a computer or digital processor so that further calculation and evaluation of the raw data is possible either during or immediately after the measurement. These systems drove the development of a complete measuring chain for the combustion pressure (also known as indication) measurement. Robust angle encoders and signal conditioning systems were also developed to provide the engine combustion engineer with a harmonised system of components for data acquisition, storage, and processing.

The most prominent feature of all true digital indication measurement devices is the ability to capture the data in the angular domain. Data are sampled at high frequency and digitised, and because the trigger for each sample is the angular crank degree mark, it is possible to sample the data at a constant rate with respect to crank angle position but with a variable frequency matched exactly to the engine speed. An overview of a typical system architecture is shown in Figure 1.14.

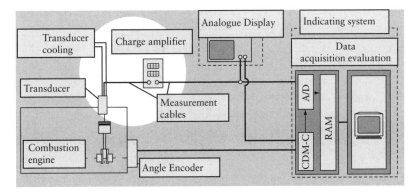

Figure 1.14 Overview diagram of a digital indication system.
(Source: AVL.)

This design ensures that the data samples can be acquired at an appropriate frequency to ensure that all the necessary harmonic components of the raw, measured curve are retained, irrespective of engine speed. In addition, the measurement frequency can be appropriately adjusted for the application, thus reducing the number of samples as necessary to optimise memory usage and system processing capability.

The first systems on the market were personal digital processor–based units. These were cumbersome units installed at the test bed, complete with customised software and user interface. Application-specific features could be included in the software such as routines to adjust the x and y axes of the pressure trace, to allow the correctly calibrated cylinder pressure curve to be displayed directly on the integrated system display screen. Data could be captured in dynamic memory and then transferred to a file format that could be stored on a hard-disk drive or transferred via optical/magnetic disk. This allowed measurement data to be processed at the test bed and also at the engineer's desk. The result was a true multi-task environment, with separation of data acquisition from analysis tasks.

As digital electronics progressed systems became smaller but also more powerful. The measurement system, including user interface, could now be moved around to wherever it was needed to execute a measurement task. Portable measurement systems based on industry-standard hardware architecture and software environments enabled these systems to be self-contained units, complete with signal conditioning (see Figure 1.15). The user could set the system up for measurement in a very short time; then, once the measurements

Figure 1.15 A portable digital indicator system.
(Source: AVL.)

had been taken, the data could be transferred via a portable medium to the test engineer or scientist, thus promoting an efficient work flow process at the test bed.

The real-time calculation ability of such systems was the most prominent step forward. Powerful signal processing capability allowed immediate analysis of the raw data to produce calculated results giving an insight into the in-cylinder process efficiency and the rate of energy release. This gave the test engineer immediate access to important facts relating to the combustion process—at the test bed, with the engine running. Thus decisions could be made regarding the test engine settings, calibration, or component selection while the engine was still available for testing. This was a major factor in the reduction of engine testing time at the test bed. It was no longer necessary to post-process the data to produce a simple set of results based on the engine pressure curve.

The availability of data derived from the pressure curve, in real time, also facilitated intelligent data reduction. The systems could reliably calculate all required results on a cycle-by-cycle basis, so that it was no longer necessary to store all the raw data curves, but only the cyclic results of interest. This amounted to a massive reduction in the amount of memory space required for a measurement process involving the acquisition of data for consecutive cycles. Real-time cyclic result calculation also allowed the transfer of result data to the test bed environment. For example, after a typical measurement procedure to obtain statistically valid data representative of specific engine conditions (for example, 50 consecutive engine cycles at fixed speed and load conditions), the data sets for each result could be processed, analysed, and expressed in statistical form (effectively giving further data reduction). These statistical data sets could be then transferred to the test bed automation system via a digital link (for example, an RS232 line), giving important log point-based information that could be combined with other measured and averaged test bed data. This provided a measurement data log with a complete overview of the engine condition, with statistically valid data values for in-cylinder combustion combined with externally measured values (temperatures, pressures, etc.) for a complete overview of the engine operating condition. An additional advantage of incorporating this level of intelligence into the combustion measurement system is that the combustion measurement process can be automated via the test bed automation system. This allows repeated measurement and data storage routines to be carried out automatically, without user or operator intervention. This reduces the tedium of repeated measurement and data storage processes that must be executed by the operator of the measurement equipment.

These features and functions have been carried through to today's equipment. Most systems now use a PC-based user interface connected to a real-time data acquisition hardware device. This allows the user to become easily familiarised with the system operation through the typical style of user interface used every day for word processing and spreadsheet applications. Using PC hardware has the advantage of standardised system architecture and components as well as a standard operating system. In addition, PC hardware is relatively low in cost and easy to upgrade. Therefore, the performance of the user interface can be easily upgraded, storage capacity can be easily extended, and the system can be readily networked via standard technologies to allow data transfer to a central host or server, to provide maximum security and easy data backup.

The real-time hardware has also benefited from technology developments. Modern systems are built from industry-standard components, and this reduces manufacturing and development costs. The performance of state-of-the-art signal processors is accelerating

Figure 1.16 A modern indicator system installed at a test bed for combustion pressure analysis.
(Source: AVL.)

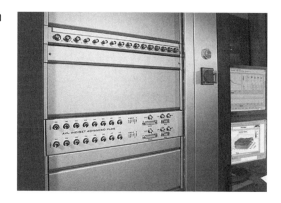

at a rapid pace, and each generation of new equipment brings new features and faster performance, with greater data storage depth and processing capability. The cost of combustion or indicator measurement systems has stabilised over the last 15 years, effectively making them cheaper but with improved performance each year. Modern systems can gather and process data quickly and efficiently; in addition, they can execute specialised algorithms for analysis methodology according to user requirements. The interface technology has improved dramatically, and current capabilities include the ability to measure, process, and share data in real time with other measurement systems and logging devices. This improves the communication process between different measurement systems and technologies found at modern test bed environments. A typical current system is shown in Figure 1.16.

The question now is, what are the next steps forward in the technology of making routine measurements to understand in-cylinder processes? As engine-related research evolves around the general goal of reduced emissions and increased efficiency, demands on the accuracy of combustion measurement will increase. Future engine technologies will require the optimisation of many more parameters that have a direct impact on the combustion event itself. In addition, control systems for these engines will require a highly deterministic combustion measurement system, to evaluate, interpret, and optimise controller settings that must be adjusted within an engine cycle and readied for the following engine cycle. The ability to acquire, calculate, and return results for use in the optimisation of high-speed engine controller control loops is a particularly demanding task for the combustion measurement system. Nevertheless, these are the forces driving the development of the systems. Data from the combustion event is essential for understanding the fundamental processes in the combustion chamber. This information is becoming more and more important and useful, and not just in the development and optimisation of engines. With future engine technologies and combustion systems, this information will be required for effective operation of the engine and its control systems. An on-board combustion measurement system in a mass-produced vehicle will be a necessity.

1.3 A Typical Measurement System

1.3.1 Complete System Components

A typical modern combustion measurement system consists of a number of component areas that are integrated to form the complete measurement chain. The process of measuring and analysing combustion pressure data occurs in a number of steps related to respective components of the measurement chain. The force applied to the measuring element in the sensor by the cylinder pressure must be converted into an electrical signal of sufficient amplitude to be recorded, digitised, and stored, ready for digital processing. Once this is done, the parameters of interest can be derived or calculated from the raw data. The steps of this conversion process are shown in Figure 1.17.

Each step in the process is summarised below.

The transducer. This converts the measured parameter of interest into an electrical signal that can be conditioned. Various technologies are employed, according to the targeted measurement. For example, for cylinder pressure measurements, the most widely employed device is the piezoelectric pressure transducer, which employs a crystal measuring element that produces an electrostatic charge as a function of mechanical force. There are other technologies available for this task, notably sensors with optical technology (discussed later in this book). For measuring other high-speed engine parameters or sub-systems, alternative sensor technologies—for example—piezo-resistive, Hall effect, and differential transformer principles—are used in an appropriate sensor package to convert pressure or displacement into a suitable signal for further conditioning and measurement.

The signal conditioning amplifier. Once the transducer produces an electrical signal that is a linear or nonlinear function of the target of measurement, this signal must be conditioned to a level suitable for high-quality digitisation and processing by the data acquisition hardware. It is important that the signal conditioning system not introduce interference or noise to the amplified signal; in addition, any inherent phase shift could lead to a serious error in the measurement system. These factors must be considered carefully. Included in the signal conditioning function are the associated connecting cables; these are used to transmit the signals and are a critical factor in providing high-quality information along the measurement chain.

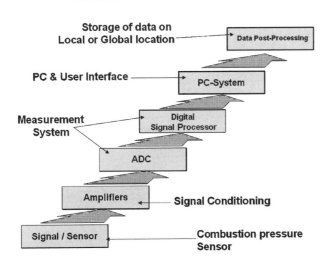

Figure 1.17 Conversion of raw data through the combustion measurement chain.

The angle encoder. The fundamental difference between a combustion pressure/indicator measurement system and other high-speed digital sampling recorders is the fact that the acquisition of data has to be sampled or converted to the true "angular" domain (as opposed to the "time" domain of, for example, an oscilloscope). The reason for this is that most of the information derived from the pressure data curve is related to crank position or cylinder volume; hence, the source data must be established accordingly. In order to do this, an angle encoder must be fitted that provides sampling marks associated with the degree of engine rotation, at an appropriate resolution. In addition, the absolute position of the engine must be established, so a once-per-revolution or once-per-cycle mark must also be provided for a reference signal.

Data acquisition system. The two fundamental, required inputs (i.e., crank angle and cylinder pressure) must be brought together to be processed, so that the parameters of interest to the engineer can be calculated, displayed, and stored for further analysis. The voltage signal supplied by the signal conditioning system is digitised by data acquisition system (sampling at crank degree intervals), where it can be subsequently stored in dynamic memory. In the process of acquisition, data are generally processed in real time so that the results can be derived from the measured curves and displayed during runtime of the measurement task. With completion of the measurement, the system allows the data to be transferred to a data file and stored permanently on a hard disk or file server. Often the system includes a personal computer as a user interface, although this is not always true of older digital systems.

There are several additional components or systems that may be utilised in conjunction with the basic measurement system components:

Dead weight tester. For effective calibration of the piezoelectric transducers and the associated measurement chain, including the data acquisition system.

Transducer conditioning system. Often included, depending on the application and the type of piezoelectric measuring element. It is needed to provide cooling fluid to the transducer element via a system that includes a reservoir, pump, and pipe work.

TDC (engine top dead centre) sensor equipment. Used for dynamic determination of TDC position of the engine. Precise TDC determination is essential, as small errors in TDC position can cause major errors in the subsequent engine calculations. The capacitive-type probe is often used, although sensors using microwave technology are also available.

It is clear that the careful choice and clearly defined integration of the working components are necessary to produce a holistic, efficient, and reliable measurement system.

FURTHER INFORMATION

1. Canadian Museum of Making, Indicators collection, www.museumofmaking.org
2. Archiving Industry, www.archivingindustry.com

Chapter 2

The Measurement Chain: Encoders

2.1 The Angle Encoder

2.1.1 Basic Function

As already mentioned, indicator measurements are referenced to crank position or cylinder volume. This concept is fundamental to the measurement principle and, therefore, all indicator systems, when used for combustion measurement, require that an angle measurement system be fitted to the engine as part of the measurement chain. An angle encoder can be fitted to the rotating crankshaft. An example of such an installation is shown in Figure 2.1. The angle encoder technologies for combustion measurement generally comprise a marker disk, an optical transmitter, and a receiver assembly. Inductive sensors and toothed wheels are also commonly utilised.

An optical system employs a light barrier to attain shine-through or reflective results. Phototransmitters and receivers are used with a marker disk to switch the light signal on and off. The transmitter generates the light signal, and the receiver diode receives it. The marker disk profile initiates the on-off switching, and the encoder electronics convert the optical signal to an electronic pulse that the data acquisition system can use. The optical sensing system determines angular displacement, using the high-precision marks on the disk, yielding high accuracy and reliability. The optical principle provides good immunity to the electrically noisy environment of the test cell and the running engine.

The Inductive measurement principle uses an engine-mounted toothed wheel in conjunction with a permanent magnet and coil. The gear's cogs pass in front of the sensor, varying the reluctance and inducing an alternating, sine current in the coil: both the frequency and amplitude of the voltage signal from this coil are functions of engine speed.

Figure 2.1 Angle Encoder mounted on the Front Pulley.
(Source: AVL.)

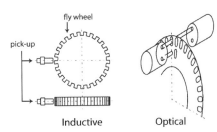

Figure 2.2 Optical and inductive measurement principle.
(Source: AVL.)

The signal processing system uses the zero crossing voltage as a reference for the cogs passing the sensor and to establish crankshaft displacement. One can also use digital Hall effect sensors with a toothed wheel to generate a square-wave signal of constant amplitude, whose frequency is dependant on engine speed. Manufacturers often use such techniques for crank position detection in electronic engine control systems. Note, though, that toothed-wheel–based sensors have limited accuracy and are generally used at low resolution (the gap between pulse edges can be anywhere from 6 to 10 crank angle degrees). Such sensors are therefore limited when used for combustion measurement, specifically to applications where crank degree resolution and accuracy are less important, for example, in engine monitoring or speed measurement. The working principles of optical and inductive sensors are shown in Figure 2.2

2.1.2 Required Resolution

The most important task the angle encoder performs is to provide referenced crank degree marks for the measurement system. As already mentioned, the data acquisition system for combustion measurement typically samples the cylinder pressure curve at regular crank degree intervals. Each sample is digitized and stored in the system's dynamic memory. The crank degree pulses from the angle encoder often initiate the sample and are thus fundamental to operation in crank degree measurement mode.

An important factor with any analogue-to-digital conversion process is to sample the measured signal at a sufficiently high frequency to ensure good-quality digital conversion and subsequent reproduction of the analogue data. According to common sampling theorems (e.g., Nyquist-Shannon), the data must be sampled at a frequency of a least twice that of the highest frequency component in the signal of interest. It is therefore recommended to sample at even higher frequencies (at 2 to 10 times the highest frequency component). If the data are digitised (a process also known as discretisation—sampling at discrete points) at an insufficient sampling frequency, then certain high frequency parts of the original signal will be lost. This is known as *aliasing* the signal.

A combustion pressure curve is a complex waveform that can consist of high frequency components of particular interest, for example, knock frequencies (in a spark ignition engine) or high frequency components that relate to noise from the combustion event (in a compression ignition engine), and it is particularly important to sample these curves at sufficiently high crank degree resolution to ensure that these components are captured successfully. An insufficient sampling resolution can completely distort the signal and affect

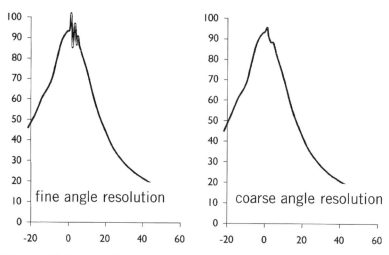

Figure 2.3 Influence of crank angle resolution on signal conversion quality.
(Source: AVL.)

the accuracy of subsequently calculated results. Figure 2.3 shows the effect of an inadequate crank angle sampling rate on a signal with a high-frequency 'knock' component.

Where a coarse angle acquisition is chosen, the knock components are completely lost. With an appropriate resolution, the pressure oscillations are completely visible.

The disadvantage of sampling at high resolution is that there are many data samples per cycle, and this sample density leads to very large data sets that can be difficult to process. In addition, there are certain applications or calculations that do not require such high-resolution sample rates. Therefore, it is important to choose a sample rate that is appropriate to the application and the measurement task at hand. Table 2.1 shows typical tasks with suggested, appropriate measurement resolutions.

Table 2.1 Measurement Task and Crank degree resolution.

Task	Resolution (Crank degrees)
Knock	<0.2
IMEP, Friction Measurements	1
Combustion Noise	<0.2
Heat Release Calculations	1
Calculation of Polytropic	1
Injection Timing Curve Edges	0.1
Direct Analysis of Pressure Curve	1
Rotational Vibration Analysis	<1

Many modern data acquisition systems designed specifically for combustion measurement incorporate a feature known as *nested* or *multilevel* measurement tables (or sampling windows). Their use allows the sampling frequency to change within the acquired engine cycle. This feature should be implemented if available because high frequency components generally occur within a specified range of crank degrees, for example, immediately after the combustion event. Thus, depending on the target of the measurement task, it may not be necessary to sample the entire cycle at high resolution if only a part of this signal of interest is at high frequency. Nested measurement allows, for example, the full cycle (360-to-360 degrees crank angle) to be measured at lower frequency (say at 1° of crank angle, for calculation of IMEP and Energy release), but it also permits a window of interest (say a 0-to-70° crank angle) to be measured at higher frequency, as this is the range where high frequency components will exist (say a 0.1° crank angle, for knock detection). Using nested measurement tables reduces the number of samples in the crank angle data set dramatically, but it still allows acquisition at an appropriate resolution where that is necessary in the engine cycle, thus conserving memory and reducing the number of data points in files and simplifying data post-processing.

2.1.3 Encoder Output Signals

The angle encoder output signal to the data acquisition system is usually a digital pulse stream. Angle encoders that generate absolute position signals in the form of a digital bit number or word are available, and they are commonly used in the process automation industry for electric machine applications because they can detect the direction of rotation. Such angle encoders are not commonly used for engine measurement, in part because of the complexity of processing the output signal, but primarily because of the lack of durability and the sensitivity of such encoders to vibration, heat, and electrical noise. The optical incremental encoder has been nearly universally adopted for accurate crankshaft displacement measurement. It provides a relative pulse output that generates an optical signal that it converts to a voltage signal.

The typical output from such an encoder system is a stream of equidistant digital pulses from marks spaced evenly around the marker disk at intervals typically from 0.1° to 6° of crank angle. This pulse train (comprising these evenly spaced marks on the marker disk) is commonly known as crank degree marks. In addition to these marks, a synchronization pulse is generated once per engine revolution. That pulse is called a trigger, or reference, mark. A typical pulse profile is shown in Figure 2.4.

The number of crank degree marks on the disk generally corresponds to a multiple of 360; most standard units generate 720 marks per revolution, that is, they are at 0.5° intervals. The marker disk is scanned optically by transmitter and receiver electronics (normally using a diode assembly). Infrared light is transmitted through a narrow lens from the transmitter diode to the marker disk. The reflected light from the mark in front of the lens is directed back to the receiver diode, where it is processed and converted.

The optical signal is converted into an electrical, square-wave signal for transmission to the data acquisition hardware. The encoder should be designed so that the processing electronics are remote from the angle encoder. Proper spacing of components in this way

Figure 2.4 Typical profile of crank degree and trigger marks.
(Source: AVL.)

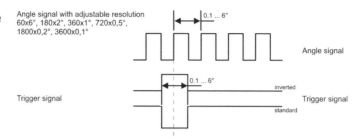

reduces the influence of engine and test bed radiated temperature and electrical noise on the encoder. The unit in Figure 2.5 portrays a design that separates the light-voltage converter from the sensor head assembly.

The raw marks generated from the encoder system can be insufficient for certain measurement applications, but multiplier electronics can be incorporated into the optical-electrical signal conditioning so additional marks can be interpolated between the raw marks to give increased resolution from the system. The use of additional marks is generally implemented via logic gates or digital signal processors using a phase-lock loop technique. Using such additional marks makes it possible to increase the resolution from a 0.5° crank angle to a 0.1° crank angle or greater.

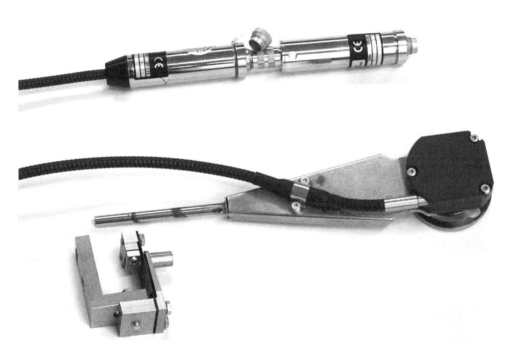

Figure 2.5 AVL 365 Encoder with separated optics and electronics from sensor head.

2.1.4 Encoder Types

2.1.4.1 Standard—Closed Encoder

The standard angle encoder for passenger car or truck combustion engine is designed for installation at the front, or free, end of the crankshaft. The encoder is mounted to the engine front pulley using a custom-made adaptor flange, and it is held in position by a suitable reaction arm to prevent torque reaction from affecting the output signal. The advantage of this design is its closed housing, which makes it less susceptible to dust and dirt contamination from the marker disk. The sealed housing contains the sensor transmitter and receiver as well as the marker disk. The marker disk itself is generally a glass disk with the mark structure applied lithographically. The optical sensor uses a reflective principle. The internal parts of a typical closed encoder head is shown in Figure 2.6.

The supporting arm contains the bearing assembly and short shaft that carries the marker disk itself at one end and the mounting flange at the other. The general arrangement is shown in Figure 2.7.

This proven encoder design is robust, and it also has the advantage that very small diameter marker discs can be used. That is helpful because errors are more dependent on the support radius than on the marker disk diameter.

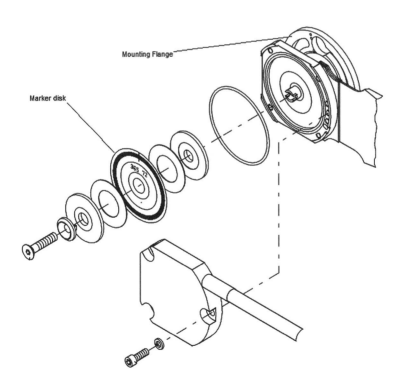

Figure 2.6 Internal Parts of a typical closed type encoder assembly.
(Source: AVL.)

Figure 2.7 Cross section of enclosed optical encoder.
(Source: AVL.)

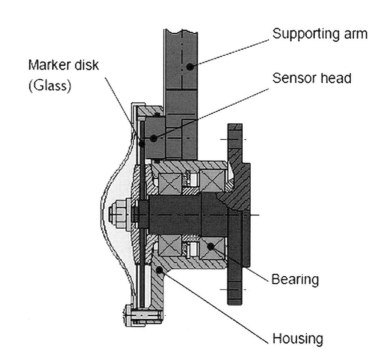

2.1.4.2 Open Encoders

Open encoders are better for certain applications. For the open design, a custom-made marker disk, specifically designed for the engine, is manufactured from spring steel. It is typically mounted at the flywheel end of the engine. The sensor head, incorporating the optical transmitter and receiver, is built into a fork assembly that mounts directly on the engine block. The open encoder uses shine-through optical measurement. The optical transmitter is fitted on one side of the sensor head. The light signal passes through slits cut into the marker disk, and the optical receiver receives it on the other side of the sensor head fork. In one design, the optical assembly comprises aligned optical fibres cast in the sensor head. There are two sets of bundled optical fibres, one for transmission and one for receiving the signal. These fibres are approximately two meters in length, and they extend to the signal processing and conditioning electronics, keeping the sensitive electronics away from engine heat and vibration.

The sensor mounting is critical because relative movements between the sensor and disk can greatly diminish the accuracy of the generated pulses. The marker disk can be manufactured with diameters typically in the ranges of 90 to 700 mm, and those diameters make such disks suitable for most engine types. The larger the diameter, the better the accuracy, although these encoders are normally used on difficult applications where a standard closed encoder cannot be fitted and where, therefore, the achievable disk diameter may be limited. A typical example of an application for the open encoder is a high-speed, high-performance engine.

Note that a typical disk is manufactured with marks at either 0.5° or 1° intervals (edges), depending on the diameter. A typical arrangement of the sensing head and disk is shown in Figure 2.8.

For very specific applications, it is possible to apply marks to an existing, rotating engine component (for example, to a clutch housing) and to use these marks in conjunction with an optical sensor. The advantage of this technique is that no additional mass or inertia is applied to the engine crankshaft. The marks are scanned using a reflective principle. Apart from that fact, the signal transmission is conventional, with the optical signal being converted to an electronic signal, amplified as required. Figure 2.9 displays a typical installation of this kind. The sensor and mark trace are visible.

In this system, an enhanced trigger mark differentiates the TDC mark from the other marks on the component. The open type encoder uses signal-conditioning electronics, producing

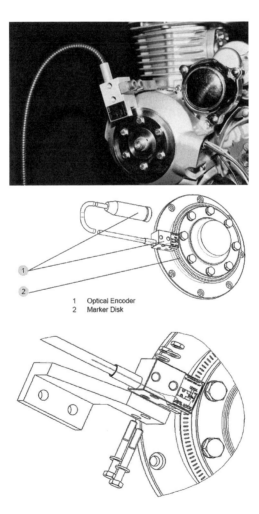

Figure 2.8 Open encoder with mounting of disk and sensor head at the engine.
(Source: AVL.)

1 Optical Encoder
2 Marker Disk

Figure 2.9 Open encoder with mark trace etched onto engine component.
(Source: AVL.)

2 separate outputs of crank degree marks and trigger marks from the single mark track. This process reduces the space required on the component for machining the marks. Figure 2.10 shows marker track detail on an engine component.

2.1.5 Processing an Existing Encoder Signal

Many engines are already fitted with toothed wheels and sensors that the engine management system uses for crankshaft position information. Although the resolution of the teeth is generally too coarse for combustion measurement, using this sensor as a source of crankshaft position information can be advantageous in certain applications, so long as reduced accuracy can be accepted. The main advantage to using such a system is the reduced

Figure 2.10 Marker track detail on engine component.
(Source: AVL.)

installation effort. Also, in certain vehicles, it may not be physically possible to mount an external encoder on the engine. An example of such an application is one for which combustion measurements are to be made in the vehicle. By this we mean an instrumented engine installed in a vehicle and driven on the road or test track to collect engine and vehicle data. That is a typical task for calibration engineers.

Generally, a toothed wheel for an engine electronic control unit generates equidistant crank degree marks between 6 and 10° of crank angle. In order to generate a reference pulse with every revolution of the engine, it is common practice to omit one tooth completely. Doing so provides a reference point on the generated pulse train that indicates to the engine control unit a specific crank degree angle (e.g., 90° before TDC) during every revolution. Common configurations are 60 teeth with 2 missing, a configuration known as 60-2. Another common configuration uses 36 teeth with 1 missing, and is known as 36-1. A pulse train of this form can be evaluated using a signal processor, and then modified to produce suitable crank degree and trigger marks to drive the combustion measurement data acquisition equipment. A typical signal from a manufacturer's engine sensor is shown in Figure 2.11.

This waveform is from an inductive sensor and it is therefore a sine form factor. The missing tooth gap can be clearly seen. The missing signal (where there is no tooth) provides a reference point. Hall effect sensors are also common in such applications. Hall effect sensors use a square wave form. Processing this signal is relatively straight forward. Here are the typical processing stages:

1. Precondition the raw signal, converting it to square wave (TTL) where necessary.

2. Supply the conditioned signal to the signal processor, which measures the time difference between incoming marks and calculates an interpolated mark interval based on the required preselected output resolution (e.g., 0.1°, or 3600 marks per revolution).

3. From this value, using multiplication logic within the signal processor, generate symmetrical output signals for the data acquisition system, with selected edge resolution, typically using an optocoupler and an open collector driver that provide TTL-level signals. (For the CDM trace, the missing tooth gap is filled with interpolated marks. For the trigger mark, the large time interval in the raw signal from the missing tooth gap is used to generate the single trigger pulse per revolution.) A typical signal conditioning process is shown in Figure 2.12.

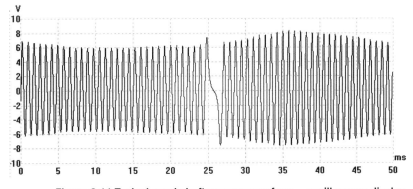

Figure 2.11 Typical crankshaft sensor waveform—oscilloscope display.

Figure 2.12 Typical signal-conditioning process of Existing Engine Crank Position sensor.
(Source: AVL.)

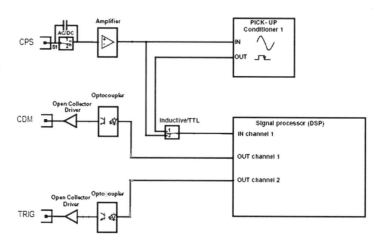

Processing an existing signal that is already available is an efficient method of gaining the required crank degree and trigger marks needed to operate the combustion measurement or data acquisition system in crank degree acquisition mode. There are, however, certain limitations to the accuracy of this method and its implementation that must be considered.

Where the source signal is generated by an inductive sensor, phase shift of the sensor signal is an issue, because with increasing engine speed, a constant phase shift causes an increasing inaccuracy of TDC positioning. That inaccuracy can have a significant impact on thermodynamic results calculated from the raw data, because they are particularly sensitive to correct TDC position. For example, a phase shift of 10 microseconds can yield a TDC error of approximately 0.3° of crank angle at an engine speed of 5000 revolutions per minute.

The output signal is generated from a number of real crank degree marks by interpolating artificial marks between the real marks, The number of input marks is generally much lower than the number of output marks, and the ratio between the two is the multiplication factor. The output marks are interpolated equidistantly by the signal processor, based on an assumed time difference between real marks. During highly transient operation of the engine (either acceleration or deceleration), or when there are torsional oscillations, a dynamic error is generated and that can lead to incorrect multiplication of marks. That is, the number of actual measured marks and the number of interpolated angle marks between two actual trigger marks does not correspond with the desired value. That situation causes an error in the output marks resulting in an erroneous time difference between the output marks. Instantaneous speed jumps appear on the pulse train measuring equipment.

The multiplication factor is critical. The lower it is, the fewer the interpolated marks between real marks. Using fewer interpolated marks reduces the possibility of incorrect multiplication in dynamic operation (as a general rule, multiplication factors less than 10 are optimum). The loading of the signal processor is a function of maximum engine speed and the required number of output marks. This formula limits the maximum achievable engine speed, based on the required resolution (according to the processor throughput rate). Engine crank position sensor signals are carefully shielded to prevent interference. In addition, they are often differential signals with no ground reference. Therefore, when one connects an external device to the sensor, one should take care to maintain appropriate shielding to protect the signal from noise pick-up. In addition, the signal can be deformed by connection of an external measuring device with a ground reference, resulting in a situation where the engine may no longer run because of corruption of the crank degree signal. Such potential effects should be carefully considered during installation of sensing equipment.

These factors must be taken into consideration when one is using measurement equipment to process an existing engine control system sensor signal. The accuracy of utilizing this sensor has measurement limitations, and accurate TDC positioning is essential (for example, where thermodynamic calculations are to be executed on the raw data pressure curves). In such cases, the limits on achievable accuracy must be carefully noted. As a general rule, this method of crank angle determination should be used only where the pressure curve, or only the results derive directly from it, are of interest.

Chapter 3
The Measurement Chain: Combustion Pressure Transducers

3.1 Cylinder Pressure Transducers

3.1.1 Introduction

There are two fundamental measurement criteria for combustion pressure measurement and evaluation—cylinder pressure and volume—and from this information further measures of cylinder work and energy release can be calculated to provide a greater understanding of in-cylinder processes. The volume, as discussed previously, is derived from the crank position which is established using an angle encoder. The installation of a pressure transducer is required for direct measurement of combustion pressure in the cylinder during the operating cycle of the engine. This is a considerable challenge: the environment in the combustion chamber is particularly harsh, with high temperatures and pressures. In addition, the transducer must be small enough to be accommodated within the cylinder head without affecting the shape of the combustion chamber, and it must be fitted to produce minimal intrusion or disturbance. This is important to avoid affecting the measurement data and to keep the costs of instrumentation within realistic limits. Another technical challenge for the transducer is the very nature of the measurement: the highly dynamic pressures in the combustion chamber have to be recorded with great accuracy and repeatability.

Selecting and installing a combustion pressure transducer always involves a compromise of technical boundary conditions and limiting factors; however, careful selection and installation, with constant consideration of the measurement task, can lead to an appropriate solution with optimised costs (in money and time).

3.1.2 Piezoelectric Pressure Transducers for Engine Combustion Measurement

3.1.2.1 Introduction

Piezoelectric transducers are by far the most commonly used sensors for combustion measurement in engine research, development, and testing. They are employed where detailed combustion analysis or combustion process monitoring is required. When used in conjunction with suitable amplifiers and means of data acquisition, they form part of a reliable and rugged measurement chain capable of highly accurate and detailed combustion pressure measurements. Generally, piezoelectric refers to the measurement technology incorporated in the sensor itself (the sensing element), which is a crystal of piezoelectric material.

The piezoelectric effect was discovered by Pierre and Jacques Curie in 1880 and refers to the property of certain crystals to exhibit electrical charges under mechanical loading. During experimentation, the Curies discovered the piezoelectric effect using a tourmaline crystal. They found that pressure applied in certain directions to opposing crystal faces produced a reverse-polarity electric charge on the surface proportional to the applied pressure. During these experiments they found that this effect also applied to other crystals—for example, quartz. This is known as the direct piezoelectric effect and is generally employed in sensors for measuring mechanical forces such as pressure and acceleration. Figure 3.1 shows the principle diagrammatically.

The reciprocal effect is shown in Figure 3.2. Here an external electrical field produces mechanical stresses in the crystal that alter its size in proportion to the strength of this

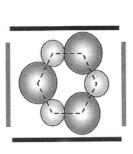
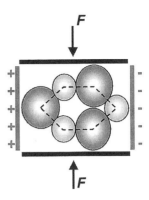

Symbolic representation of a piezoelectric crystal for the example of quartz. The silicon atoms are shown as large and the oxygen atoms as small balls.

Charge displacement with a force acting on the piezoelectric element.

Figure 3.1 Piezoelectric crystal charge effect.
(Source: Kistler.)

Direct piezoelectric effect	**Reciprocal piezoelectric effect**
A mechanical deformation of a piezoelectric body causes a change in the electric polarisation that is proportional to the deformation.	An external electrical Field E causes mechanical stresses proportional to the field, which alter the size of the piezo-crystal.
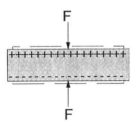	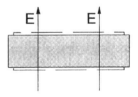
For measuring mechanical parameters, especially of forces, pressures and accelerations	In ultrasonic and telecommunications engineering

Figure 3.2 Direct and reciprocal piezoelectric effects.
(Source: AVL.)

field. This effect is used in ultrasonic and communications engineering. One of the first applications for the technology was during wartime, in sonar devices. The reciprocal effect has an automotive application in the form of piezoelectric operated fuel injectors in modern, common-rail diesel fuel injection systems. In this case, the element is housed in the injector body and is charged electrically to produce the required force and displacement to open the fuel injection orifice, supplying an accurately measured amount of fuel to the cylinder for the combustion event. The advantage of this method is the extremely low inertia of moving parts compared to the use of a solenoid-type valve, and this allows operation of the fuel injector at very high speed, facilitating the incorporation of multiple injections required for noise and emission reduction.

The direct piezoelectric effect was of no real use in measurement technology until very-high-input-impedance electronic amplifiers enabled engineers to condition the signals. In the 1950s electronic components of sufficient quality became available and the piezoelectric effect became usable for practical measurements. The charge amplifier principle was patented by Kistler in the 1950s and gained applicability in the 1960s. The introduction of appropriate semiconductor technology (MOSFET—metal oxide semiconductor field effect transistor) and the development of highly insulating materials (such as Teflon and Kapton) greatly improved performance and promoted the use of piezoelectric sensor technology in many areas of pressure and force measurement in modern research, development, and industrial applications.

The operating principle of the combustion pressure transducer is the fact that the piezoelectric sensing element within it generates an electrical charge in proportion to the stress applied to it. The in-cylinder gas pressure is applied to a diaphragm in the sensor body that is exposed to the combustion chamber. As pressure changes are applied to the diaphragm, it in turn applies a dynamic force on the measuring element crystal, generating the charge signal as a function of the cylinder pressure. Note that this sensing technique is "active." That is, a charge is produced only when the force on the crystal is changing; hence the output from the sensing element is a function of the change in stress or force on the measuring element (the first differential of the force applied), and therefore a piezoelectric sensor cannot detect true static pressures. Note, though, that it is possible, with an appropriate signal conditioning system, to perform quasi-static measurement over very long periods of time.

For combustion measuring applications, dynamic pressure is the main target. However, for correct positioning of the pressure curve to enable the accurate determination of absolute values, the static pressure must be established at some point in the engine cycle for referencing of the signal. This can be done in conjunction with an absolute measuring sensor or, as is most common, the offset can be calculated empirically from the dynamic pressure measurement data on a cycle-by-cycle basis (commonly known as zero-level correction or pegging).

3.1.2.2 Crystal Materials Used

The material most commonly associated with piezoelectric measurement is the quartz crystal (SiO_2). Although there are other piezoelectrical materials available, quartz is one of the most appropriate for combustion measurement, as other materials are subject

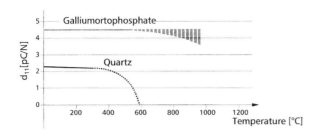

Figure 3.3 Temperature dependency of piezoelectric materials.
(Source: AVL.)

to compromises that can affect the quality of the measurement data in a combustion measurement application. Examples are:

Tourmaline and langasite: Charge output is affected by temperature.

Lithium niobate and tantalate: Resistance decreases as temperature increases.

Piezoceramics: Temperature-dependent effects are a problem in the material.

Most of these adverse affects are directly related to the operating temperature. Even with quartz material, the temperature effects are a significant issue. Figure 3.3 shows the effect of temperature on the piezoelectric constant for quartz material.

Thus, as a material for cylinder pressure transducers, quartz is still very temperature-sensitive; in addition, the limit with respect to maximum operating temperature of the measuring element is generally around 250 °C. This puts specific limits on the operational temperature range such that in general, for a simple crystal element transducer to be used for combustion measurement, a cooled type of transducer, plus an appropriate cooling system, must be selected. This configuration allows the measuring element in the transducer to operate reliably and repeatably, even at the high temperatures, which may be encountered at the transducer-engine interface, depending on the mounting position of the transducer in the combustion chamber.

An alternative method of extending the temperature range of the quartz element is to optimise the orientation of the crystal cut. The crystal elements used in the sensor can be cut to optimise charge output as a function of the applied force and stresses in the element. For combustion measuring, elements with longitudinal or transverse crystal cuts are generally used; these are shown in Figure 3.4.

The distinction between these two crystal cuts is the position of the charge on the element relative to the applied force. The Kistler Company (www.kistler.com) employs a "polystable" cut for certain sensors, which is optimised to reduce the temperature effects on the sensitivity of the measuring element. With this technique, the piezoelectric behaviour of the element remains constant within a defined temperature range, with a maximum operating temperature of around 350 °C. This allows the use of a practical, uncooled sensor for combustion measurement applications.

There are also synthetic crystal technologies that have been developed specifically for use in high-temperature applications for piezoelectric sensors. Gallium orthophosphate ($GaPO_4$)

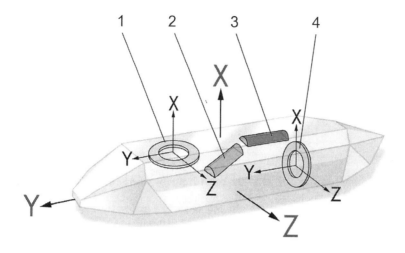

1 Cut for the longitudial effect
2 Polystable cut
3 Cut for the transverse effect
4 Cut for the shear effect

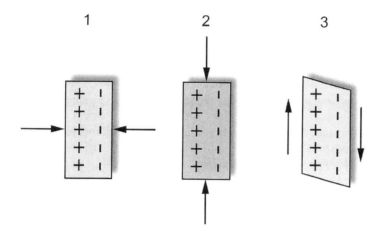

Crystal cuts and piezoelectric effects: longitudinal, transverse, and shear effect.
1 Longitudinal effect 3 Shear effect
2 Transverse effect

Figure 3.4 Crystal cuts, longitudinal and transverse.
(Source: Kistler.)

was developed and is used exclusively by the company Piezocryst (www.piezocryst.com) in their combustion measuring sensors. This material has a particularly high piezoelectric sensitivity that is nearly independent of temperature. The main advantages of this material compared to quartz are:

- Temperature resistance up to 900 °C
- High sensitivity compared to quartz (nearly double)
- High insulation resistance (discussed later)
- No effect of temperature on charge output

This material has proven particularly appropriate for the development of smaller transducers for modern production engines, where the availability of space in the combustion chamber to accommodate a sensor installation has become a technical challenge. In addition, this material has facilitated the development of alternative, adaptor-type pressure measuring solutions that allow nonintrusive access into the combustion chamber via existing in-cylinder engine components (i.e., glow and spark plug ports). Sensors with both the quartz and $GaPO_4$ elements, are available and due to the crystal technology employed, are uncooled. That is, even with no cooling water infrastructure for the transducer, the performance of the quartz or $GaPO_4$ element compares with that of a cooled transducer. This reduces the installation cost and the complexity of in-cylinder access; in addition, the problems associated with transducer cooling systems (e.g., failure leading to overheating, cross-talk of pressure pulsations on the measured signal) are eliminated.

3.1.2.3 Transducer Construction and Types

A very wide range of transducer designs are available for combustion measuring applications. The basic design depends on the engine size and application, but the general form and construction are similar for most types. A cross-section of a typical sensor is shown in Figure 3.5.

Generally, the measuring element is housed inside the transducer body, which allows installation, sealing, and heat transfer away from the sensor. At the front end of the sensor, exposed to the cylinder pressure, is the diaphragm. Pressure applied to the diaphragm produces a proportional force that is applied to the measuring element. This force creates the stress in the piezoelectric element that generates the charge signal. In most cases the measuring element comprises a "stack," or multiple elements to increase the charge output from the sensor. This signal is brought out to an external connector for transfer via a measuring cable to the signal-conditioning amplifier. Depending on the sensor, this signal could be ground referenced or ground isolated, as shown in Figure 3.6.

Ground isolation is used to increase immunity to electrical noise in the test environment. This can be a worthwhile feature, but it should be noted that ground-isolated sensors must be used with compatible amplifiers. Note also that the sealing arrangement of the sensor is an important factor. The sensor installation requires precision machining around the mating surfaces; any leakage of hot combustion gases around the sensor body will cause premature failure of the sensor due to excessive heat load. Thus, this interface is a critical area.

Figure 3.5 Schematic drawing of a typical piezoelectric sensor.
(Source: AVL.)

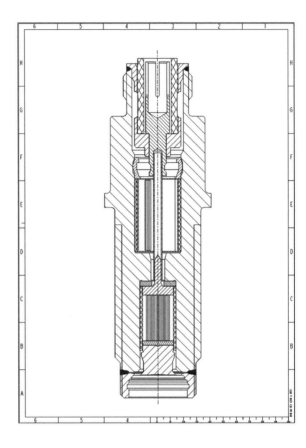

Cooled Type

The traditional cooled type of transducer is the most commonly used and trusted transducer for accurate and detailed combustion analysis and measurement tasks. Generally, this is the largest type of transducer available in thread mounting sizes from M10 to M14. Due to the larger size, they can accommodate a correspondingly larger measuring element, and

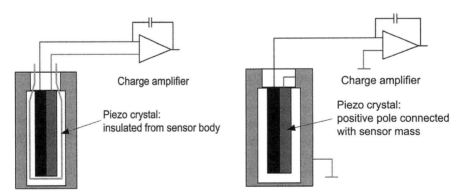

Figure 3.6 Ground-isolated and insulated sensor principles.
(Source: Kistler.)

this provides a relatively high charge output relative to the applied pressure. The general perception is that, due to water cooling, they are more stable and less sensitive to errors introduced by thermal conditions. However, with the modern crystal technology used in uncooled sensors, this is a misconception.

Applications for water-cooled transducers are limited due to restrictions in installation space in modern engines of smaller displacement (for example, passenger car engines). However, they are widely favoured for use in detailed combustion research carried out in single-cylinder research engines, where the cylinder head normally has fewer constraints for accommodating a transducer installation. In addition, water-cooled sensors can be utilised where small pressure changes are to be measured—for example, in manifold pressure dynamics, where the signal amplitude is relatively small and a sensitive sensor with a high output is needed. The cross-section of a typical cooled sensor is shown in Figure 3.7.

Note that this sensor has the measuring elements arranged in a stack, utilising the longitudinal effect cut, which results in the production of a sensor with high charge output and optimised signal-to-noise ratio. It is also possible to construct transducers of this type with transverse-cut crystal elements as shown in Figure 3.8.

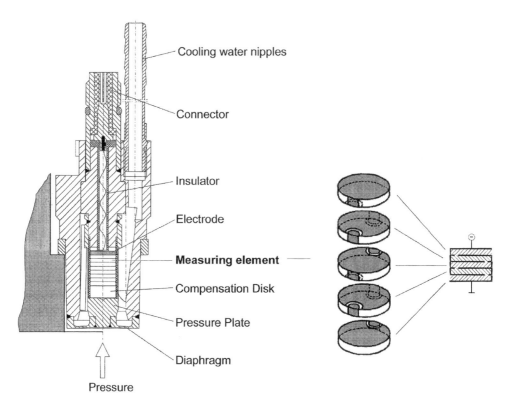

Figure 3.7 Cross-section of a cooled sensor with stack elements.
(Source: AVL.)

Figure 3.8 Cross-section of a cooled sensor with transverse-cut elements.
(Source: Kistler.)

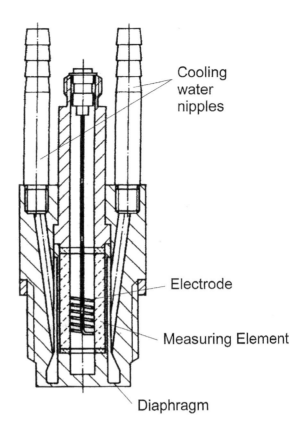

The water flows in the cooling jacket around the measuring element and diaphragm, transferring heat away during operation. This ensures that the measuring element temperature remains stable and within operational limits. In fact, the measuring element temperature is only slightly (approximately 10 to 20 °C) above that of the cooling water.

The cooling system uses de-ionised water in a tank with pump, manifold, and appropriate tubing to supply and return the coolant. The use of de-ionised water is essential to prevent the buildup of calcium deposits, which can block the cooling channels in the transducer over time. A stable, pressure-pulse-free flow of coolant to the transducer is essential, as pressure pulsations can be picked up by the measuring element and superimposed on the measured signal (a phenomenon known as cross-talk).

A typical arrangement is shown in Figure 3.9.

Uncooled Type

Uncooled sensor technology has improved dramatically in recent years. The introduction of new crystal technologies and improved manufacturing techniques has allowed the production of small sensors, with acceptable sensitivity and stability levels, suitable for most common combustion measurement applications in nearly all engines. The accuracy and

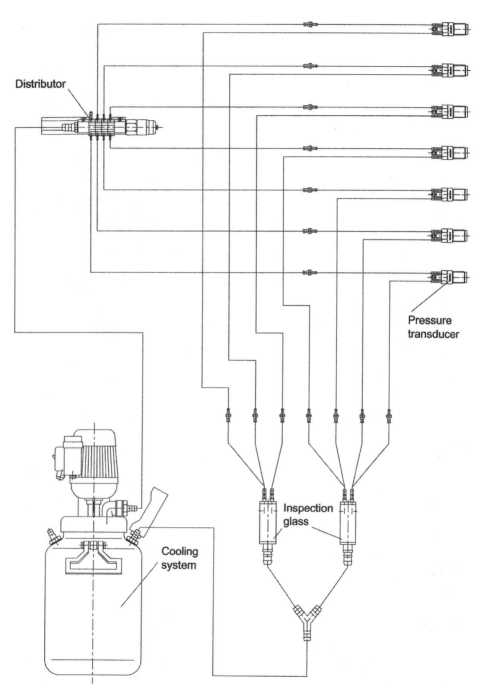

Figure 3.9 Cooling system overview.
(Source: AVL.)

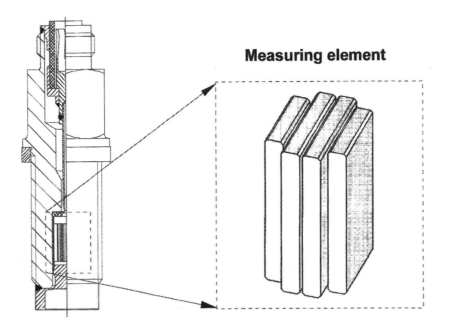

Figure 3.10 Uncooled transducer, cutaway view.
(Source: AVL.)

sensitivity of current sensors in this category are now comparable to the levels of equivalent cooled sensors, but with the advantages offered by the lack of requirement for a cooling system. In addition, the small size of sensor elements in this class allows the use of advanced adaptor technology, allowing access into the cylinder with minimal intrusion and thus the possibility of combustion measurement where it typically has not been possible in the past (for example, in-vehicle on the road).

Typically, uncooled sensors come in 5-mm threaded or 6/7-mm threaded or plug-in mounting types. In addition, the compact measuring elements can be installed in glow plug housings to allow a measuring glow plug to replace the working glow plug when access into the cylinder is required. Because uncooled sensors require high stability and temperature resistance of the measuring element, only optimised, synthetic crystal materials are used in transducers of this type. Figure 3.10 shows the structure of a typical miniature pressure transducer.

In this design, transverse-cut measuring elements are used that help to keep the measuring element compact. Generally, due to their smaller size, uncooled transducers have a higher natural frequency and are thus more suitable for applications where higher-frequency measured components are involved. Typical examples are measurement of combustion knock or noise, and measurements at high engine speeds. The smallest threaded transducers are often used in conjunction with adaptors for nonintrusive access to the cylinder. The slightly larger threaded and plug-in types are used with adaptors for permanent installation in the cylinder head.

Small uncooled sensors are now available with sufficient sensitivity that they can be used where detailed thermodynamic analysis is required. However, it should be noted that greater changes in sensitivity can be expected at higher operating temperatures, irrespective of the crystal technology employed, often due to construction-related factors (e.g., diaphragm deformation). For this reason uncooled transducers are characterised by a higher cyclic temperature drift (i.e., temperature drift within an engine cycle) and a higher load change drift (temperature drift over a number of cycles).

3.1.2.4 Piezoelectric Transducer Properties

Introduction

The piezoelectric transducer for combustion measurement has to operate in an extremely harsh environment. The effects of this environment on the measuring system and the subsequent data quality must be fully understood. In addition, the various characteristics of a piezoelectric sensor must be considered prior to selection and installation of the transducer for the measurement application.

The basic properties of the transducer defines its effectiveness in producing a signal relative to the cylinder pressure, with more complex parameters determining its performance and signal quality with respect to the environmental conditions that it is forced to endure in operation. These are particularly important to understand and consider when a detailed analysis of the combustion pressure curve is necessary.

Environment Effects

The transducer encounters a number of environmental effects once installed in the engine cylinder:

Temperature. The transducer is exposed to extremely high gas temperatures at relatively high pressures, under very dynamic conditions. Consequently, the transducer can face extremely high heat flow loads in operation. Actual temperatures vary, but a transducer may have to operate at temperatures up to 500 °C at the front area of the diaphragm, with gas temperatures in the cylinder greater than 2000 °C.

Accelerations and vibration. Structure-borne vibrations created by the engine's reciprocating parts cause accelerations at the pressure transducer that can be superimposed on the measured data curve. Accelerations of 1000 g are commonly encountered, and in extreme cases, such as high-performance engine applications, the acceleration values can peak at 2000 g.

Deformation. The forces generated by the gas pressure in the cylinder, in addition to the thermal stresses due to expansion and contraction, can generate large stress forces of up to 200 N/mm^2 in the cylinder head material at the transducer mounting interface. These stresses can be imposed on the transducer body and have a distorting effect on the measured value.

Deposits and contamination. Inside the cylinder, carbon or other deposits can build up on the transducer face, causing corrosion or affecting the measurement quality. Externally, dirt or some other contaminant can affect the quality of the output on the signal transmission line. Figure 3.11 illustrates these external influences on the transducer's efficiency.

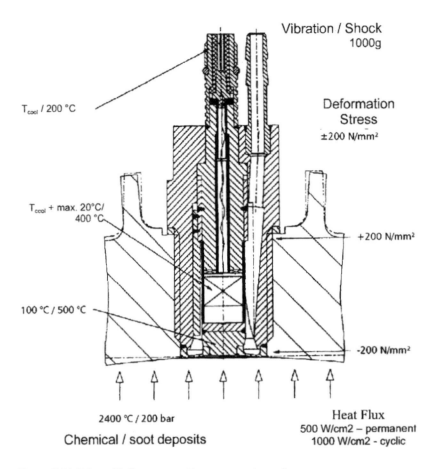

Figure 3.11 External influences on the pressure transducer.
(Source: AVL.)

Transducer Properties and Specifications

Predicting or assessing the response of the transducer to external influences involves consideration of a number of important parameters relating to the properties and performance of the piezoelectric transducer. In certain cases, specific test methods are called for in order to specify the performance level. The main parameters of interest are as follows.

Measurement range. The operational pressure range of the transducer. It is the range over which the transducer is guaranteed to perform within the stated tolerance and to meet the specifications on the data sheet with respect to sensitivity and linearity, for example.

Overload capability. If the pressure measurement exceeds the normal upper limit, then the overload range becomes a relevant concern. Most transducers can withstand some degree of overload without suffering any irreversible damage; however, in this range the accuracy of the transducer is not guaranteed. However, in most cases, the data and the transducer remain usable (see Figure 3.12).

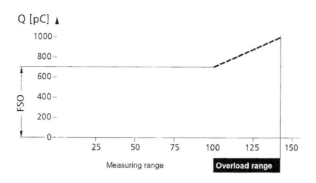

Figure 3.12 Measurement and overload range.
(Source: AVL.)

Operating temperature range. The temperature of the sensor has a significant effect on the sensitivity of the measuring element, so the operating temperature range must be monitored. Outside this range accuracy is compromised. Many transducers offer sensitivity as a function of specified calibration range and temperature range on the calibration certificate supplied with the unit. This parameter is most important where applications with large changes in temperature are to be expected—for example, where engine cold-start and warm-up operation measurements are to be executed.

Operating life. The transducer has a finite lifetime that is highly dependent on the operating conditions it is exposed to and the quality of installation. This property is defined as the number of load cycles (or engine cycles) over which the transducer retains its technical performance properties. Note that operation under highly loaded conditions dramatically shortens the projected life of the transducer. Typical or average operating conditions are used by transducer manufacturers as a basis for projecting life; a generally quoted figure for the life of a unit is approximately 10^7 operating cycles. An important means of optimising the life of any transducer is the use of dummy units. Even if the transducer is not being used for measurement, if it is fitted in the engine and subjected to pressure cycles, it is subject to wear and its remaining life is decreasing (irrespective of whether the engine is firing or not). An appropriate way to optimise transducer life is to remove the unit from the engine every time (when practical) the engine needs to be run but no measurement is required (for example, during warm-up or conditioning of the engine) and replace it with a transducer dummy, which is a transducer-shaped unit that seals the engine bore. These are generally available from transducer manufacturers; an alternative is to use an expended transducer as a dummy, presuming it is physically intact.

Sensitivity. Sensitivity can be defined as the change in signal output from the transducer, relative to the change in the pressure applied to it. Generally, transducers are marketed by the manufacturer with a nominal (or average) sensitivity value. All transducers show a certain amount of sensitivity scatter over the operating range, and for this reason every sensor is supplied with a complete and individualized calibration sheet that covers the operating range and states actual values derived from that transducer, measured during factory calibration before supply to market. It is important to use the suggested calibration value for the operating conditions of the transducer in the application (with respect to temperature and pressure). Generally, for piezoelectric transducers the calibration is stated in picocoulombs (unit of charge) per bar/psi (unit of pressure).

Linearity. The transducer uses a piezoelectric measuring element that produces a charge relative to the applied force that is a function of the pressure at the transducer diaphragm. However, certain deviations from ideal performance, as a result of manufacturing and production tolerances, can occur across the operating range; this is known as linearity error. It is generally specified as a percentage of the full scale output (% FSO, i.e., maximum for the stated operation range of the transducer). Linearity is derived from the calibration curve using the best straight line with forced zero point (BSLFZ), as shown in Figure 3.13.

The mean sensitivity is defined as the gradient of the BSLFZ. This "best straight line" is the centre straight line passing through zero between two parallel lines that capture the pressure transducer characteristic as closely as possible. The distance between the two lines that contain and represent the characteristic of the transducer is then expressed as a percentage of the full-scale value:

$$\text{Linearity}[\%] = \pm \frac{A}{Q_{max}} \cdot 100$$

If certain operational pressure ranges are of interest during the measuring task (e.g., for peak pressure measurement), then the most accurate measurements—those with the lowest

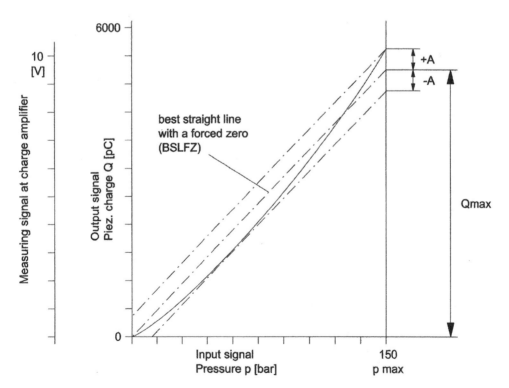

Figure 3.13 Transducer sensitivity and linearity.
(Source: AVL.)

deviations from the ideal—are obtained via an optimised calibration of the transducer, executed within that pressure range only.

Temperature coefficient of sensitivity. This parameter describes the reversible change in sensitivity due to deviations in the temperature of the transducer (within the specified operating temperature range) from the temperature at which it was calibrated. It is expressed as the actual change in sensitivity, as a percentage of the nominal or mean sensitivity of the transducer, per °C (within the operating temperature range).

In practice, water-cooled transducers do not undergo wide temperature variations at the vicinity of the measuring g element; therefore, the changes in sensitivity are negligible and can be largely ignored. If certain operating temperatures or extremes of temperature are likely to be encountered, then appropriate calibration of the transducer at the specific temperature can be executed, and this will reduce or negate the effect of temperature sensitivity shift on the measured data.

Natural frequency. The natural frequency is the frequency of free vibration oscillation of the complete transducer. Generally, for systems with minimal damping (as is the case with pressure transducers), the natural frequency of the first order is equal to the basic resonant frequency.

The basic resonant frequency is the frequency at which the pressure transducer delivers the output signal with the highest amplitude. The natural frequency of the transducer must be sufficiently high that oscillations from the measured signal do not excite the transducer into resonance, as this would cause false high-frequency components to be superimposed on the measured data curve. As a general rule, the highest frequency component expected in the measured signal should not exceed 20% of the resonant frequency of the transducer.

Insulation resistance. The insulation resistance is the resistance measured between the output signal connector of the transducer and the transducer housing (which is connected to ground). High electrical resistance is imperative in the piezoelectric measuring chain, as too little resistance can cause charge leakage and electrical drift. This parameter is important not only for transducers but for the complete system, including cables, connector adaptors, and charge amplifiers. As a general rule, the resistance value of the complete chain should be at least 10^{13} ohms at room temperature.

Note that transducers can be not only ground referenced but also ground isolated; here the measuring element is completely isolated from the transducer housing and connected externally via terminals. Ground-isolated transducers can prevent the passage of ground loop currents, which are due to differences in ground potential between the various distributed component parts of the measuring system. Ground-isolated sensors must be used with the correct amplifier equipment but are suitable for use in electrically sensitive measurement environments, as they can provide good signal quality without interference suppression.

Capacitance. Due to their construction, piezoelectric pressure transducers possess some capacitance from the electrodes of the measuring element and the natural capacitances of the connector and lead. Typically in combustion measuring applications, charge amplifiers are used; in this case these capacitances have no significant effect on the measuring chain as long as the measuring cable between the transducer and the charge amplifier is not excessively long (i.e., less than 20 metres). If cables longer than this are used, the cable capacitance could be noticeable by means of a slight increase in noise and reduced bandwidth of the charge amplifier.

Acceleration sensitivity. Sensitivity of the transducer to structure-borne vibrations that can cause false frequency components to be imposed on the measured signal. This factor is normally specified by the manufacturer of the transducer in units of bar/g. The actual effect of strong mechanical vibrations on the measurement depends on the mounting position of the transducer. Note that mounting transducers on exhausts and intake manifolds can be problematic because of this effect due to the lower mass of these components compared to that of the cylinder head. High-performance or "racing" engines suffer particularly from the effects of structure-borne noise on the quality of the measured pressure signal because of their high engine speeds and loads.

Cooled sensors generally have higher sensitivity to acceleration forces because the dynamic response of the transducer is influenced by the mass "resonance" of the cooling water both in the pressure transducer and in the connecting hoses. The sensitivity is therefore usually considerably higher than in transducers without cooling. It is commonplace for acceleration sensitivity to be specified in the technical data tables for cooled transducers for operation both with and without cooling water.

Transducer manufacturers have developed "acceleration-compensated" units in order to reduce the effect of structure-borne vibrations on the measured signal. The basic principle is to use an additional measuring element in conjunction with a seismic mass along with the pressure-measuring element. The sensitivity of the additional element is matched to the pressure-sensing element but is of the opposite polarity. The two signals from the elements are combined, canceling out signals generated by the tuned vibration frequencies, and this provides a high degree of desensitization to these vibrations—typically about 0.1 mbar/g. The principle is shown diagrammatically in Figure 3.14.

The Kistler Company (www.kistler.com) has developed this concept further via a passive-compensation technique (as opposed to the active concept just described). This technique does not require any additional compensating measurement elements and is therefore ideally suited to small, uncooled transducers. Inside the transducer, the measuring element is supported by a sleeve that itself forms a dynamic spring-mass system. This mounting system for the measuring element is tuned to the same natural frequency as the measuring diaphragm of the transducer; hence the measuring element is effectively "sandwiched" between the two spring-mass systems (diaphragm and mounting sleeve). In this way, if an acceleration acts on the transducer, both the diaphragm and the mounting-support sleeve system will respond by oscillating with the same amplitude, in the same direction. This ensures that the measuring element receives no additional force due to the vibrations and that the signal produced is free of these effects. The achievable compensation using this method is typically 0.2 to 0.4 mbar/g. Figure 3.15 shows the basic principle of "passive" compensation. The graph shows measured data from a high-performance engine; the advantages of using acceleration-compensated sensors in this application can be clearly seen from the measured curves.

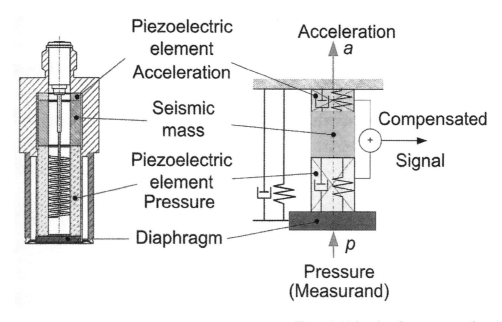

Figure 3.14 Acceleration compensation.
(Source: Kistler.)

Torque (mounting). This is simply the torque applied to the transducer when it is mounted in the installation bore (assuming it is a threaded mounting). It will always be quoted in the specification sheet for the transducer, and it is imperative that the correct figure be observed to ensure correct sealing at the front measuring face. This is necessary to prevent gas leakage, which can affect the readings and the response of the engine. In addition, hot gases leaking past the measuring face and mounting threads will overheat and destroy the transducer in a very short time.

The transducer is carefully calibrated at a specific mounting torque. The mounting torque stresses the body of the transducer and measuring element and therefore has a direct

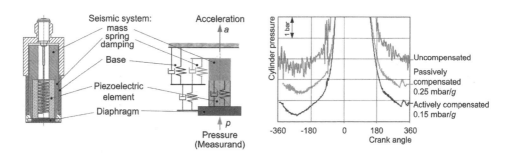

Figure 3.15 "Passive" acceleration compensation.
(Source: Kistler.)

consequence on the sensitivity of the unit. If the mounting torque is applied carefully, the change in sensitivity due to this effect is negligible. Use of the correct mounting torque is particularly important for small sensors with limited sealing surfaces; improper torque can lead to major pressure differences at the sealing surface interface and cause leakage or sensor damage.

Transducer Thermodynamic Properties

Thermodynamic characteristics are particularly important in assessing a transducer's suitability for extremely accurate measurements and further calculation or analysis of the engine pressure curve.

As previously mentioned, transducers are sensitive to temperature, and any deviations from the temperature at which the transducer was calibrated that are encountered during use can be a source of error that must be accounted for. For combustion measurements, this temperature sensitivity can cause signal drift, that is, a deviation in the measured pressure that is due purely to temperature changes at the transducer and the mounting position.

The non-steady-state heating of the pressure transducer exposed to the cylinder combustion gases causes this drift effect, and the amplitude and time characteristic of the temperature-related drift are functions of the pressure transducer and installation position heat flow. It is particularly important to understand which transducer surfaces are heated and, to a lesser extent, how the transducer mounting arrangement deforms when heated.

In order to properly assess and quantify the quality and performance of a pressure transducer under dynamic conditions with respect to temperature drift, it is necessary to measure its performance under precisely defined conditions of pressure, heating, and installation position. From this information, the performance data must be converted into meaningful and objective characteristics for users to properly compare the performance of given units under the specific operating conditions of the test and the application.

These temperature-related sources of signal drift can be divided into two categories: drift effect within and over cycles:

Within a cycle (thermal shock, short-term drift). This is the drift error that occurs within a single engine cycle and is due to the cyclic heating of the transducer. It is caused by thermal stresses in the diaphragm induced by the heat flow from the combustion gases, which is of short duration but high intensity. The sensing diaphragm expands and contracts due to the heat flow, and this causes the force on the measuring element to differ from that applied by the cylinder pressure. The problem is worst at low engine speeds due to the longer cycle time (and time for heating effects to develop). Other factors can also affect the amount of drift, such as the mounting position of the transducer and the engine operating conditions—start of energy conversion point, speed, load. The cyclic drift affects all of the parameters derived from the pressure curve—the most significant being IMEP, which can be subject to errors of up to 10%.

The effect is represented in Figure 3.16, which shows the deformation of the diaphragm under heat load as calculated via a FEM simulation.

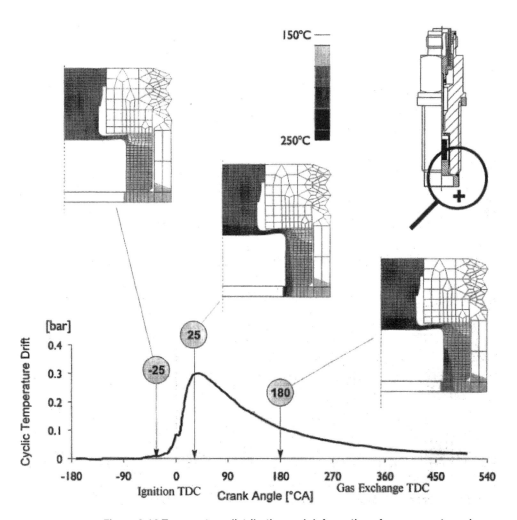

Figure 3.16 Temperature distribution and deformation of a pressure transducer.
(Source: AVL.)

There are two methods for determining the performance of a transducer with respect to cyclic drift. The first is by comparison with a reference sensor in a fired engine. Alternatively, a cyclic heating device can be used for the determination, a method that applies no actual pressure on the transducer itself. When the comparison method is used, the transducer is installed in a test engine alongside the so-called reference transducer, which is a transducer specially selected with respect to linearity and drift behaviour. It is silicon coated and has thermal protection to reduce temperature effects; often the plug-in type is used to reduce deformation errors, and it may also be water cooled and/or use thermally stable measuring crystal elements. The sensor is cleaned regularly and has a short service life, and for these reasons it can be relied on to give accurate and repeatable results as a baseline for comparison with production sensors.

Figure 3.17 Short-term drift effect on a sensor output compared to a reference sensor signal.
(Source: Kistler.)

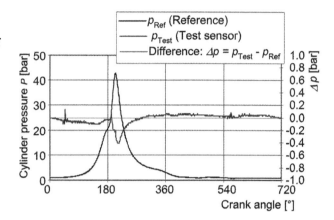

The simplest evaluation involves comparing the measured curve from the test sensor with the reference sensor under fixed operating conditions of speed (often quite low—around 1500 revs/min) and load (often quite high, at IMEP ~ 10bar). The maximum pressure deviation within a measured cycle in relation to the pressure value before the start of energy conversion is the defined characteristic. A typical curve is shown in Figure 3.17.

With modern transducer technology, cyclic drift is optimised and can be so low that it is difficult to establish or measure objectively in a running engine. This is due to the fact that the cyclic heating effects can be so small that they are difficult to discriminate from other interference sources. For this reason, cyclic heating test rigs can be used to assess the performance of the transducer with respect to such effects. A typical test rig is shown in Figure 3.18.

Figure 3.18 Cyclic temperature drift test rig.
(Source: AVL.)

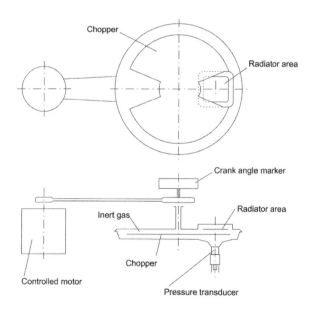

The test rig unit applies a heating source, representative of that found in an engine installation, to be applied cyclically to the transducer at a frequency analogous to that in the engine. The unit consists of a radiant heat surface that is positioned opposite the pressure transducer under test. The heat flux from this source is alternately concealed and exposed by a rotating chopper wheel, and this results in a heating curve as illustrated in Figure 3.19.

The pressure transducer is flush-mounted in a solid, cooled steel plate and is exposed periodically to an electrical heat source that is over 2000 ºC and can produce heat flow of more than 100 W/cm^2 to the transducer under test. The maximum pressure deviation that occurs in a cycle (relative to the measured signal immediately prior to the heating phase) at a specific frequency of heating, for a specified heat flow, is the characteristic value used to describe the cyclic drift behaviour.

Further calculations regarding the absolute performance of the pressure transducer with respect to intra- and intercycle effects are possible. For example, the dynamic sensitivity of the transducer can be established in order to calculate the error caused by cyclic temperature drift alone in relation to the total error (see Figure 3.20).

Establishing the dynamic sensitivity helps to identify the true error that can be apportioned to cyclic drift phenomena, as a simple comparison of the pressure difference error between reference and test sensors includes errors due to temperature and deformation.

This dynamic sensitivity value can then be utilised to correct IMEP drift over cycles in order to separate and understand the long-term drift effects of short- and long-term temperature change, and mechanically related effects on the measurement data. It is important to fully understand these effects and trends of the pressure transducer drift-related characteristics and their interrelationships because, under certain conditions, one error can be opposite in polarity to another, making the net effect on the measurement appear negligible. However, as the measurement conditions (i.e., engine operation mode, speed, and load) change, these errors can accumulate at certain points to produce significant unwanted effects. It is clear that, in general, a low sensitivity to thermal shock is favourable for accurate calculation of results derived from the pressure curve.

Temperature effects over cycles (long-term drift, load change drift, gas exchange drift). An engine operates in transient conditions of speed and load, and changes in these conditions have an effect on the temperature and heat flow load experienced by transducer; this in turn has consequences for the transducer signal, over a number of cycles, after a change in operating conditions. It manifests as a slow variation in the pressure signal after a load change and is due to the altered thermal stresses in the transducer body, which have an effect on the forces acting upon the measuring element. It continues to have an effect on the measured curve and cycles until the mean temperature of the transducer stabilises after the change in operating conditions. It is worthwhile to note that this phenomenon is more noticeable in uncooled transducers.

The characteristic values to express this phenomenon are determined in a real engine, under fired operating conditions. The engine is operated at a specific speed and load condition,

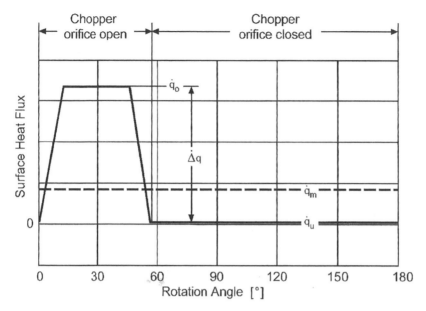

Figure 3.19 Heat flow to the transducer in the cyclic temperature drift test rig.
(Source: AVL.)

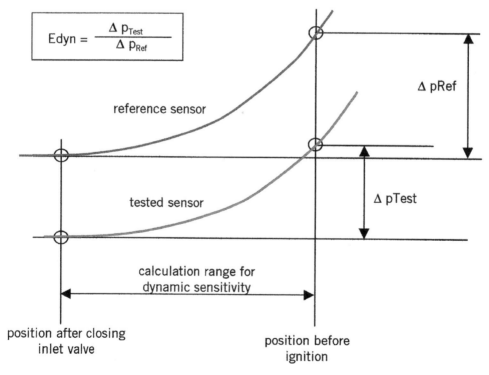

Figure 3.20 Dynamic sensitivity.
(Source: AVL.)

then a step change in heat and load is effected by switching instantaneously to motoring mode; this is implemented via fuel supply shutoff. Figure 3.21 shows the effect.

In the diagram, the combustion cycles before and after a load change can be seen with a change in maximum pressure from 42 bar down to 14 bar. The temperature of the transducer drops due to the change of heat flux; this has the effect of causing deformations that effectively "unload" the transducer's measuring element. Referring to the lower curve, which is formed by the minimum points of the cycles, these zero-points when joined into a single line show the zero-line curve trend and its deviation from the ideal zero-line. From these data, two meaningful results that represent the long-term drift characteristic can be derived in order to compare transducers for suitability. The first of these is the zero-line gradient, which represents the change in pressure level over time caused by the change in heat flow load. The second characteristic is the permanent deviation of the zero-line, normally measured 20 seconds after the load change.

Of these two characteristics, the former is the most significant as it has a direct effect on the measured curve within a cycle, and hence affects calculations derived from that curve (similar to other intracycle effects). Therefore, for accurate IMEP and heat release

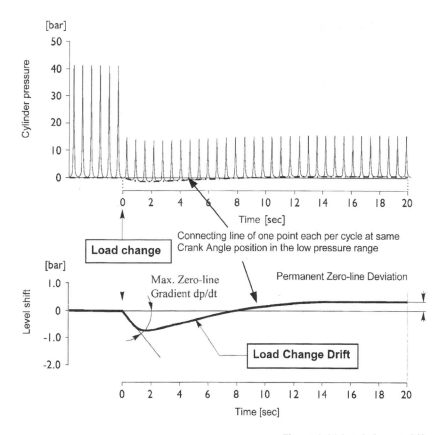

Figure 3.21 Load change drift.
(Source: AVL.)

calculations this property is particularly significant, especially where measurements are to be made during transient conditions.

The latter effect is a permanent zero-level offset of the measured curve; it has a lesser effect on the measured data, and with modern amplifiers equipped with drift compensation circuitry, this effect is minimised such that it can be considered insignificant.

IMEP stability. Depending on the engine, fuel, and operating condition, deposits caused by combustion residues and lubricants that are normally present in an engine can have an adverse effect on transducer accuracy over a period of time. This is particularly important where transducers are in use for extended measuring periods; the deposits built up on the diaphragm face and the annular gap between the transducer and bore can affect the rigidity of the diaphragm and the heat flow at the transducer. Thus, these factors directly influence the sensitivity, the linearity, and above all the cyclic temperature drift of the transducer and therefore affect transducer performance and calculation from the measured data.

This effect can be characterised by IMEP stability, which is the percentage change of IMEP over a defined run time relative to a reference transducer. To measure this effect, the test transducer is installed in an engine with the reference transducer; they are operated simultaneously and the IMEP results are compared to determine the IMEP stability. During this procedure the reference transducer is cleaned and calibrated at regular intervals to ensure the stability of the reference conditions. At the end of the test run, the test transducer is also cleaned to restore its original condition. A typical test run to determine IMEP stability is shown diagrammatically in Figure 3.22.

IMEP stability gives a clear indication of the sensitivity of a transducer to soot build-up. This can be particularly important for some applications—for example, where measurements must be made in the exhaust system using cooled transducers.

Summary: Transducer Design

The transducer has to operate in harsh conditions, and a number of factors can affect the stability and repeatability of the sensor output as a function of input. Comparison of a particular transducer with a reference transducer can facilitate understanding of the behaviour of that unit for a particular application. Some aspects of transducer behaviour cannot be determined with certainty, however, due to occurrence of errors under certain engine speeds, loads, and energy conversion conditions relative to a reference transducer mounted in the same cylinder but in a different position.

Many of the causes of inaccuracies or imprecise behaviour can be controlled by careful design of the transducer itself. Linearity has little effect, but thermal and dynamic influences can lead to a significant shift in the behaviour of the sensor. Care in the choice and application of the crystal used as the measuring element can optimise the transducer's response to temperature change. Advanced crystal technology and optimised crystal orientation in combination with housing design can reduce thermal sensitivity shift effects considerably.

Dynamic sensitivity changes are related to deformation of the transducer housing and membrane due to stresses at the mounting bore. This type of error is proportional to the applied pressure and can be reduced via the use of plug-in mounting sensors or double-shell

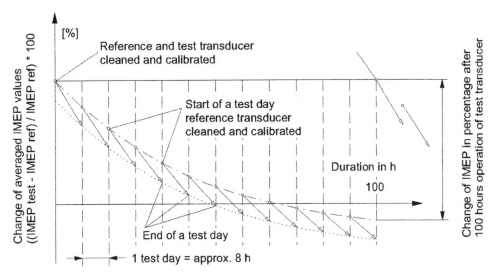

Figure 3.22 Test to determine IMEP reproducibility and stability.
(Source: AVL.)

transducer bodies. These constructions limit the interaction between the measuring element and its surroundings. The degree of dynamic sensitivity shift is related to characteristics of the mounting bore, adaptor, and mounting position. Therefore, the precise effect for a specific installation can be very difficult to predict.

Good thermodynamic performance is critical, and this depends heavily on the choice of design to reduce temperature-related effects both within an engine cycle and over a number of engine cycles. Intracycle drift (cyclic drift) can be reduced by optimising the design and geometry of the measuring membrane. Intercycle drift (load or gas exchange drift) can be reduced through careful design of the transducer body to ensure minimal, or at least predictable, thermal expansion of the body parts so that pre-load changes to these parts due to temperature have little effect on the output of the measuring element.

The overall goal of transducer manufacture is to ensure the best possible behaviour in response to any of the preceding influences under any operational or mounting conditions. It is important to note that in some cases errors can accumulate but also offset each other. For this reason, a good understanding of the characteristics of piezoelectric measurement technology is important in choosing the proper transducer for a particular application.

3.1.2.5 Transducer Installation and Adaptors

Introduction

The method used for mounting the transducer for access to the combustion chamber has a major effect on the accuracy of the measurement and the lifetime of the transducer itself.

There are two main avenues for gaining access: via the measuring bore or via an adaptor inserted in place of a standard component. These methods are known as intrusive and nonintrusive means of access, respectively, and each has its advantages and disadvantages. The choice of access method depends ultimately on the measuring task, the evaluation, the time necessary for the required modifications, and the costs involved. Figure 3.23 shows the relative merits of the two methods.

Intrusive Mounting

An intrusive installation is one that involves intervention via precision modification of the engine parts (cylinder head) so that the transducer and measuring face are suitably positioned for exposure to the gas pressure inside the combustion chamber. This type of adaption has many variations due to the large number of possibilities with respect to installation position. In addition, the complex interdependencies of the boundary conditions that cause error—for example, excessive heat loads—must be observed.

Generally, intrusive mounting can be executed with or without an adaptor. An adaptor sleeve is required when access to the combustion chamber requires traversing oil or water passages. An additional advantage of the adaptor sleeve is that it effectively separates the transducer body from the surrounding cylinder head material, thus isolating it from deformation stresses that could causes a shift in sensitivity during engine operation. It should be noted that direct installation can be effected simply and that this method allows easy removal and installation of the transducer.

Figure 3.24 shows a direct-mounted transducer employing front sealing, which should be used whenever possible. With this method of sealing, a smaller surface area of the transducer is exposed to the hot gases and heat flow and the sensitivity shift due to deformation is reduced.

Transducer Mounting Considerations and Positions

The exact position for the transducer depends on a number of interrelated factors, but the position chosen has a considerable influence on the final result with respect to data quality. There are a number of factors that must be considered carefully. In a modern engine, transfer passages for oil or water could block the route between the external environment and the chamber. The passages must remain open following the transducer installation to prevent a change in the thermal conditions in the cylinder. Engine components in the

Figure 3.23 Comparison of adaption with and without intervention.
(Source: AVL.)

Adaptation with and without intervention in the test engine		
	Intervention in test engine	No intervention in engine
Installation time and effort	substantial	little
Costs	high	low
Possible to select measuring position	yes	no
Accuracy	high	medium high

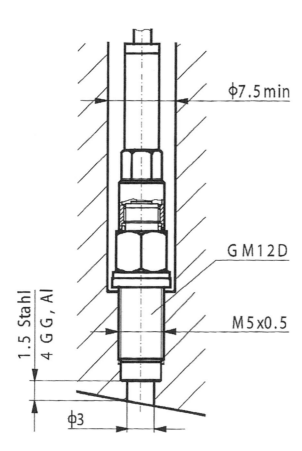

Figure 3.24 Transducer using front sealing and direct installation.
(Source: AVL.)

combustion space—such as valves, injectors, and spark plugs—must also be unaffected by the installation. Structural integrity of the engine parts must also be preserved, and the wall thicknesses of casting must not be reduced to the point where weaknesses could cause premature failure.

Ideally, the measuring face is positioned nearly flush with the surface of the combustion chamber, recessed just enough to prevent deposit build-up on the piston from damaging the transducer, but not so much as to create a measuring pipe that can produce high-frequency oscillations at resonance or increase cylinder volume and thereby reduce the compression ratio. In certain instances, slightly recessing the measuring face reduces the thermal load, and this can increase the transducer's life. As a general rule, the recess must not be greater than the pipe diameter.

The pressure in the cylinder varies in relation to the measuring position due to the gas dynamics, for example, around squish areas. Therefore the actual position chosen for installation must provide a representative value according to the target of the measuring task. In addition, errors can be introduced because of environmental factors, such as heat flow, temperature, and accelerations, so all these boundary conditions must be considered carefully when choosing the installation site.

Gas dynamics during the normal cylinder filling and exiting processes can affect the measurement, although locating the measuring face near the inlet valves helps to reduce heat flow load. High gas flow rates near the valve openings can affect the pressure measurement by causing regional pressure differences; should this occur, the measured pressure is not representative of the combustion chamber pressure. This can be a problem particularly during the gas exchange part of the engine cycle. Mounting the transducer adjacent to an exhaust valves will also affect the performance due to temperature-related issues; the area around the exhaust ports is subjected to the highest temperature, and the hot, fast-moving exhaust gases flowing over the measuring face of the transducer will significantly increase heat flow load.

Inclination of the access point to the cylinder also has a negative effect. Generally, the transducer's longitudinal axis should be at a right angle to the cylinder surface; if this is not possible a small inclination angle can be tolerated, but anything greater will increase the combustion chamber volume, risking interference to the gas flow conditions in the cylinder. This could cause fuel droplets to collect around the installation point, which can have a dramatic effect on emissions measurements.

In certain adaptor configurations the measuring face must be recessed for practical reasons (to reduce heat flow load and temperature) and in these cases an indicating channel is created. As mentioned previously, a pipe or channel between the sensor face and the combustion chamber produces undesirable effects that must be accounted for during installation and measurement. The additional volume can change the compression ratio, thus affecting operation of the engine and its performance. In addition, this volume can trap fuel particles, affecting emissions. The indicating channel also acts as a resonator with a natural frequency that can be excited by the combustion pressure activity, causing a high-frequency disturbance on the measured pressure curve that can be difficult to distinguish from combustion chamber oscillations. The effect can be estimated by modeling the system as a simple Helmholtz resonator as shown in Figure 3.25.

The resulting effect is shown in Figure 3.26. The gas temperature curves are plotted as a function of frequency and indicating channel length. The diagram is derived from a typical installation of a 5-mm front-sealed transducer.

In addition to the oscillation effects, long indicating channels cause phase shift in the pressure signal and can increase temperature due to high flow rates at the diaphragm.

There are some specific considerations with regard to the mounting position that are related to engine type (gasoline or diesel). For gasoline knock measuring applications, the position of the transducer measuring face has a decisive effect on the measurement of the knock modes during detonation or pre-ignition. The knock phenomenon is caused by high-frequency pressure waves; under certain circumstances these can be reflected from the wall of the combustion chamber and cause a surge wave that registers as a high-amplitude knock at the transducer, should the transducer lie in the path of this wave at the cylinder periphery. This produces misleading pressure readings (compared to the cylinder average) with high-frequency and high-amplitude signal wave components. To reduce this effect, the transducer can be centrally mounted in the cylinder. However, it should be noted that a transducer mounted in this central position may not detect the first, fundamental knock

$$f = \frac{\sqrt{\kappa R T}}{2} \sqrt{\frac{r^2}{V\, l}}$$

κ .... Isentropic exponent
R .... Gas constant [J/kg K]
T .... Gas temperature [K]
c_V .... Specific thermal capacity at constant volume [J/kg K]
r .... Radius of the indicating channel [m]
l .... Length of the oscillating gas column (in indicating channel) [m]
V .... Volume [m³]

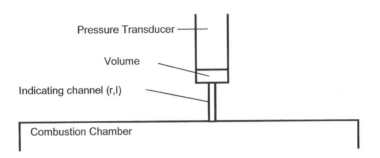

Figure 3.25 An indicating channel.
(Source: AVL.)

mode, which is generally the highest in amplitude and easiest to detect. Therefore, the knock detection algorithm must be carefully selected and utilised in order to correctly establish the limit of detonation during engine mapping or tuning.

There are similar challenges and compromises in diesel engine applications. Modern diesel engines are generally of the direct injection type. This means that a sensor mounted in the combustion chamber will either be in a squish region or above the piston bowl; both

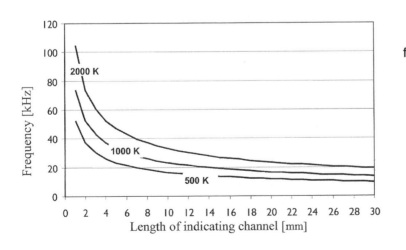

Figure 3.26 Pipe oscillation as a function of channel length and gas temperature.
(Source: AVL.)

positions have their advantages and compromises. Sensors mounted over the piston bowl respond favourably to the applied pressure, giving an output signal with no delay. However, they undergo high thermal loading and hence, when transducer longevity is important (e.g., engine monitoring), special care should be applied in mounting the transducer to protect the measuring face from the flame radiation.

Sensors mounted in the squish area can produce distorted pressure readings due to gas dynamics generated by the motion of the piston around TDC; however, thermal loading is reduced, and for monitoring applications, this position is ideal. It should be noted, though, that the transducer must be recessed to prevent deposit build-up on the piston, which could damage the transducer membrane.

Consideration of all these factors is essential in the selection of the measuring position, and careful attention to all of these points will ensure high-quality measurements with predictable compromises where these have to be made.

Installation of Mounting Bores

Once the transducer type, mounting position, and method have been selected, the practical task of engineering the means of access into the combustion chamber must be undertaken. This should be considered a precision engineering task and should be executed with care and forethought to ensure a quality installation that will provides representative, good-quality data.

Specialised machining tools are available, and these should be employed during processing of the engine components for direct or indirect (via adaptor) transducer installations.

Direct installation involves drilling of a suitable passage using the appropriately stepped drill (drill a centring hole first and then mill the entry point to allow for the mounting tool). Once this has been correctly and accurately drilled, the thread can be cut using the appropriate tap. On completion, the bore and thread should be de-burred and cleaned thoroughly before installation of the transducer. Note that for miniature, un-cooled sensors it is generally not required to first drill a centring hole, due to the small diameter. In addition, the machining tolerances are reduced as the transducer size is reduced. Figure 3.27 shows the stages of a direct installation.

Adaptor or mounting sleeves are employed where direct installation is not possible because internal media passages in the cylinder head have to be traversed. When installing a mounting sleeve, the main considerations are the following:

- The flow in any media passages must not be affected by the bore position.
- Structural integrity must be maintained.
- The sleeve must be leak proof, reliable and not affect the engine conditions or operation.

In order to achieve these conditions, it is generally necessary to carry out a detailed analysis of the cylinder head, using engineering drawings to evaluate the options for locating the transducer in the cylinder. This means that sleeve installation is generally more time-

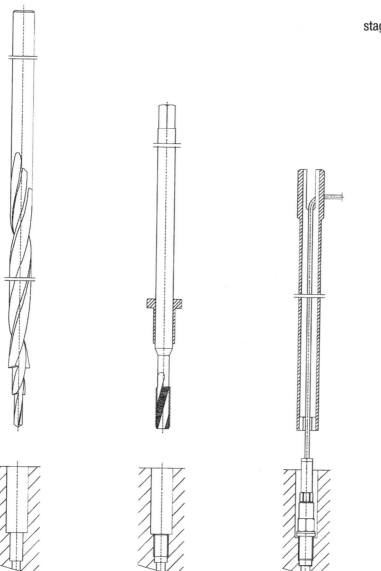

Figure 3.27 Work stages for transducer installation.
(Source: AVL.)

consuming than the direct installation technique. The machining process requires high precision in order to be reliable, and it is essential that the correct machining tools are employed for the process. Figure 3.28 shows a typical installation of an uncooled transducer using a sleeve.

Installation of the measuring sleeve must allow for expansion and contraction of the materials due to temperature change; this may be achieved using O-rings or elastic bonding agents. Whichever method is used, the appropriate machining tolerance must be observed,

The Measurement Chain: Combustion Pressure Transducers

Figure 3.28 Example installation: uncooled transducer with adaptor sleeve.
(Source: AVL.)

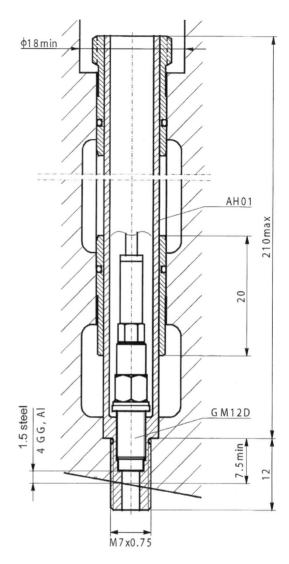

to ensure no leakage. On completion of the installation, any protruding part of the sleeve can be cut off and machined flush to maintain the surface profile and quality on the combustion chamber side. The final step is to ensure that the installation does not leak (done in water bath), and to ensure all passages are free of machine swarf and cleaned ready for re-installation and use.

Note that adaptor sleeves can be relatively simple single-sleeve installations or more complex installations where multiple sealing points are required. An additional complication in the machining of the installation is created if the bore enters the cylinder at an angle (as opposed to perpendicular). In this case the complexity and number of work stages involved

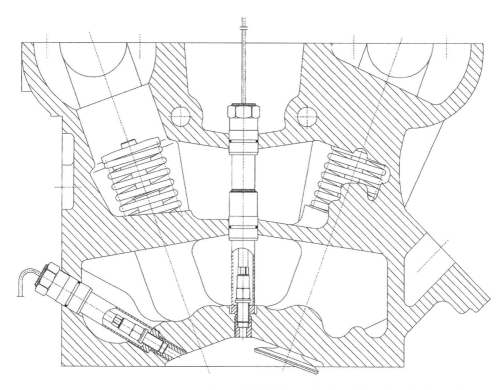

Figure 3.29 Cylinder head with two installed transducers.
(Source: AVL.)

is increased. Figure 3.29 shows a cylinder head with two transducer installations, one single-sleeve but inclined, one perpendicular but with multiple sealing points.

Nonintrusive Mounting

As crystal technology has developed, particularly with the introduction of miniature, uncooled transducers, it has become feasible to produce a transducer and adaptor assembly of sufficiently compact size that could replace an existing component which has access to the combustion chamber and thus could be installed and removed easily. In general terms, this means spark plug adaptors for spark ignition (gasoline) engines or glow plug adaptors for compression ignition (diesel) engines.

Clearly, the advantage is that the amount of work to gain access into the combustion chamber is greatly reduced. It is simply a case of removing and replacing a component in the cylinder head, as would be carried out in an engine service. This has a significant effect on the cost and time required to instrument the engine.

The disadvantage is that there is limited choice regarding where the measuring face of the transducer can be placed in the cylinder. It has to be positioned at the location of the component that the transducer assembly is replacing; thus, the accuracy or quality of the measured data could be compromised in some way. However, the above advantages are attractive, and for certain applications the compromise can be accepted as long as it is fully understood and quantifiable.

Nonintrusive access to the cylinder is commonly used in engine monitoring applications, where detailed analysis of the pressure curve (i.e., indirect/thermodynamic analysis) is not carried out, as direct analysis of the curve is more important in these applications. In addition, nonintrusive access is common for durability, or calibration work, where the engine is less likely to be equipped with a measuring bore (as it will be closer to a production design engine).

Another common application for nonintrusive access is for in-vehicle work (often engine control system calibration tasks). In this case, the ease of instrumenting an engine installed in a vehicle is attractive to reduce work time preparing and making measurements with the engine installed in the vehicle.

It is worth noting that developments and evolution of nonintrusive adaptor solutions have improved the available solutions over the years and reduced the compromises in the quality of the measurement data, to such an extent that today nonintrusive solutions are a practical alternative that, in most cases, provide data of quality that is nearly comparable with an intrusive solution—and hence, they can be specified as a suitable, cost-effective alternative for thermodynamic analysis of the pressure curve in some applications.

This trend is likely to develop further as pressure transducer miniaturisation continues. In addition, another driving factor is the likely adoption of in-vehicle combustion measuring for production vehicle engine control systems. Next-generation combustion technologies and engine control systems need feedback on the progress and quality of the combustion process, to be utilised in high-speed control loops that monitor engine fuelling and emissions. This will drive the development of crystals and sensors for use in nonintrusive, instrumented engine components (i.e., spark and glow plugs), for feedback and monitoring purposes in production engines.

Spark Plug Adaptors

For spark-ignited engines, the nonintrusive method of accessing the cylinder is via spark plug adaptors. These units are instrumented spark plugs that perform a dual function. They introduce the electric spark in the cylinder to ignite the mixture, as would a normal spark plug. In addition, they house and locate a miniature pressure transducer in an appropriate position, such that the measuring face of the transducer is exposed to the combustion chamber pressure. Also, the spark plug adaptor must be easy to remove from the engine and, when installed, must provide an appropriate gas-tight seal. The adaptor must be designed to be as close as possible to the original equipment with respect to heat range and position of the electrical arc. This is particularly important in modern, direct-injection gasoline engines, as the stratified mixture must be ignited in the area where the mixture strength is sufficiently rich to initiate combustion. Any displacement of the spark position can have a decisive effect on the quality of the ignition process. An additional technical challenge is the environment that the sensor has to operate in. Not only with respect to heat but, due to the electrical high voltage, the spark plug adaptor must separate

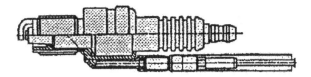

Figure 3.30 First-generation spark plug adaptor.
(Source: AVL.)

the high voltage and charge signal, preventing cross-talk and interference of the signals and, in addition, maintaining the required level of insulation resistance to ensure no leakage of the high-voltage spark that could cause misfire, even under the most extreme engine operating conditions.

These are considerable technical demands. Nevertheless, as the crystal and sensor technology has developed, the design of the spark plug adaptor has evolved to bring the measuring face closer into the cylinder, with better signal quality available from it. Figure 3.30 shows the first generation of spark plug adaptors. This was the original equipment spark plug, adapted via the installation of a suitable thread in a "piggyback" position, adjacent to the insulator, where a miniature transducer could be installed. The measuring face is connected to the cylinder via a gas channel formed in the thread of the spark plug. The advantage of this design is that the original equipment spark plug is used; hence, heat range and spark position are entirely as the manufacturer specifies. The disadvantage is the very long gas channel, with consequently low acoustic resonance and detrimental effect on signal quality.

A development of this type was an improved version, a specifically manufactured, measuring spark plug (as opposed to an adapted, original unit). The design is shown in Figure 3.31. It incorporated an inclined electrode position that allowed the transducer face to be located closer to the cylinder. This provided a shorter gas channel between the transducer and the cylinder, with a consequent increase in the measurement quality achievable. This allows the detection of high-frequency components for knock measurement and monitoring. The main disadvantage of this design was that the indicating channel still exists between the transducer face and the combustion chamber.

Although both of the above types have the disadvantages associated with an indicating channel, the position of the transducer (remote from the combustion chamber) means that

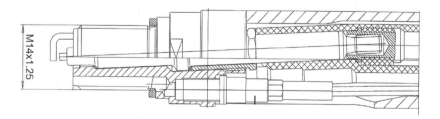

Figure 3.31 Measuring spark plug with inclined electrode.
(Source: AVL.)

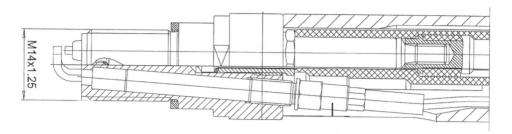

Figure 3.32 Measuring spark plug with probe.
(Source: AVL.)

the heat flow load on the sensor is reduced, and hence short-term (intra-) cycle sensitivity shift is much less. As crystal technology developed, though, the opportunity of bringing the measuring face closer to the cylinder became possible, and the next generation of spark plug adaptors incorporated miniature measuring probes installed in adaptor housings that allowed direct access into the combustion chamber with virtually no indicating channel. Figure 3.32 shows a typical adaptor of this type.

This arrangement puts considerable demands on the transducer and adaptor design. The measuring probe and element must be able to withstand the in-cylinder conditions. In addition, the mechanical strength and integrity of the adaptor must be maintained, to provide the necessary gas-tight seal and to be durable in use. The transducer probe must also be removable from the adaptor to allow servicing and replacement. Improvements in the crystal used have increased the sensitivity of the transducers for this solution, such that the signal-noise ratio is optimised, thus providing a good-quality signal. In addition, this spark plug adaptor has nearly no indicating channel, in combination with high natural frequency, making it ideal for detection of detonation and knock. The ease of adaption for cylinder access provided by this solution makes it attractive for engine knock calibration and tuning applications. Note, though, that the centric position of the spark plug in the cylinder (common with modern engines) means that first modes of knock cannot easily be detected, so alternative algorithms that do not rely on this first mode must be used for successful detection of knock limits.

Also, it is worth noting that with this solution, the electrode position has to be shifted off-center to allow the transducer installation. This has a negative effect in certain engine types where spark position is critical, and this should always be considered.

Current developments of the spark plug adaptor include improvements in the design with respect to the installation. Early types relied on the use of insulator extensions to raise the spark plug cap (incorporating the electrical connector) away from the ceramic body to allow access for connection of the measuring cable. Under certain circumstances this can cause problems with electrical noise from the ignition system affecting other components (due to reduced noise and electromagnetic suppression). Additionally, this exposure of the ceramic can cause leakage of the high-voltage spark, with the consequence of engine misfiring (very

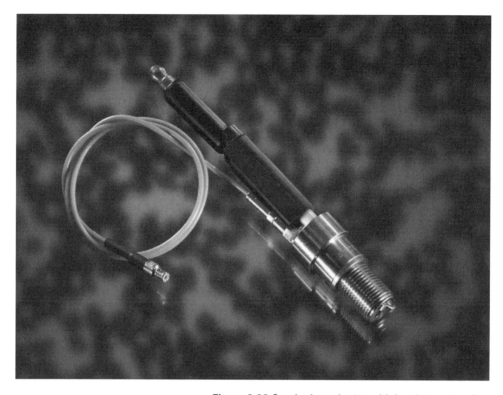

Figure 3.33 Spark plug adaptor with insulator extension.
(Source: AVL.)

undesirable during testing). A typical arrangement with an insulator extension is shown in Figure 3.33.

Some current designs of measuring spark plugs and spark plug adaptors have very compact, integrated measuring elements or sensors that allow exiting of the cable directly from the spark plug body. This means that no insulator extension is needed; thus no disturbance of the high-tension electrical connection is required. This improves the reliability of the installation. Additionally, this method of adaption is now easily accommodated with the alternative ignition coil arrangements found on modern vehicles (coil packs and pencil coils mounted directly on the spark plug). A typical design is shown in Figure 3.34.

Note that in most cases, the spark plug adaptor or sensor is supplied in a number of standard heat ranges. It is important to correctly identify the heat value according to the manufacturer of the spark plug. Confusingly, several manufacturers have different conventions with respect to heat value numbers. The table in Figure 3.35 shows some examples of these correlations among manufacturers.

There are several crucially important factors to be noted when using spark plug adaptors:

The spark position, geometry, and heat value of the adaptor must match the original equipment as fitted. Any deviations will affect the quality of the data measured, and this

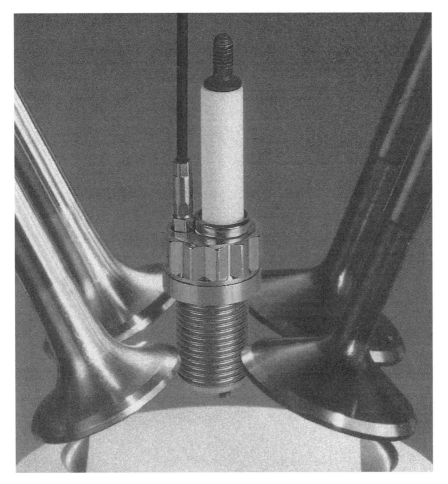

Figure 3.34 Measuring Spark Plug.
(Source: Kistler.)

should always be considered and understood. Most manufacturers of spark plug adaptors can provide detailed advice and can instrument spark plugs on a bespoke basis for nonstandard applications.

The spark plug adaptor, by its very nature, has no degree of freedom with respect to mounting position of the measuring face of the transducer in the combustion chamber. Therefore, for certain applications, the use of these adaptors can be limited. As mentioned previously, some knock modes are not detectable by the spark plug–mounted transducer, although this can be overcome via alternative knock algorithms. It should be stated though that spark plug–mounted transducers are generally not suitable for measurements that require accurate determination of absolute values from indirect results—for example, IMEP. This is normally required where accurate, absolute determination of cylinder contributions or energy balance is required. However, for general monitoring and day-to-day measurement tasks, spark plug adaptors are suitable and appropriate.

DENSO	NGK	Champion	BOSCH
9	2	18	10
14	4	16, 14	9
16	5	12, 11	8
20	6	10, 9	7, 6
22	7	8, 7	5
24	8	6, 61, 63	4
27	9	4, 59	3
29	9,5	57	
31	10	55	2
32	10,5	53	
34	11		09
35	11,5		07

Figure 3.35 Spark plug heat range comparison table.

When installing spark plug adaptors, the correct mounting tools should always be used, and the correct mounting torques must be observed. In addition, always check that the engine has sufficient space around the spark plug seating and installation area to ensure that the installation can be effected successfully.

As technology proceeds, the design of spark plug adaptors has evolved and will evolve further. Currently the trend in gasoline engines is toward smaller-diameter spark plugs. This reduction in diameter presents challenges for manufactures of the adaptors. M14 and M12 spark plugs are common for passenger car engines with displacements of around 0.5 litre per cylinder. However, the current trend is to downsize engines, and this means that smaller spark plugs, of around M10 size, will become standard for these engine applications. Note that current designs of spark plug adaptors at M12 and M10 sizes (at the time of writing) cannot accommodate a transducer measuring face close to the cylinder with undisturbed arc position. The most likely scenario will be a measuring spark plug that has an integrated sensor element, close to the combustion chamber. It is expected that with the trend in combustion measurement techniques, optical as well as pressure-sensing spark plug adaptors (simultaneous acquisition) will be needed. These will provide a better understanding of the in-cylinder dynamic processes for the development of future combustion engine technologies.

Glow Plug Adaptors

Glow plug adaptors are the nonintrusive access solution for compression ignition or diesel engines. Most engines of this type are fitted with preheating probes. These can be removed and replaced with an adaptor that contains a suitable, uncooled miniature pressure transducer, and this can access the combustion chamber via the probe body.

Often the measuring task for a modern diesel engine involves the establishment of the combustion noise level. This is the noise contribution provided by the combustion event alone and is a result derived from the cylinder pressure curve. With modern diesel engine injection technology, multiple injection events can reduce the high-intensity, short-duration energy release in the cylinder (that provides the characteristic diesel knock), and often the combustion noise level is an optimisation target when engine mapping. Of course, glow plug adaptors offer convenient access into the cylinder to measure and derive this parameter. The only issue is that an indicating channel in the glow plug adaptor can be resonated by the combustion noise generating pressure rises. For this reason, most designs of transducer/adaptor are now optimised such that the transducer face is as close to the combustion chamber as possible. This reduces the pipe length between the measuring face and probe tip, thus reducing the pipe oscillation effect. Whether this pipe oscillation is visible on the pressure trace depends upon the sound pressure level of the engine at the resonant frequency of the glow plug adaptor (defined by the length of the indicating channel). Any resonance effects will artificially amplify the combustion noise level, and engines with high noise (i.e., sound pressure levels) will require adaptors with correspondingly high resonant frequencies (i.e., short indicating channels). This cannot always be fulfilled with a glow plug adaptor design, and in these cases, a measurement bore must be installed as an alternative.

Modern engine designs are particularly challenging, as the glow plugs are often very slim (~4mm), and the installation of a transducer inside an adaptor with appropriate dimensions is impractical. Either the transducer would have to be mounted far from the tip of the probe via an indicating channel (with associated oscillation effects induced), or the adaptor walls would be so thin that they would have no mechanical strength. In this case the solution is a measuring probe—that is, a transducer in the shape of a glow plug. This arrangement allows a sufficiently slim design, with the measuring element located at the front of the probe, close to the combustion chamber. The only disadvantage of this design is that the transducer cannot be replaced. Once its accuracy and repeatability behaviour become outside usable limits, the whole probe must be replaced. A typical sensor/adaptor and measuring probe are shown in Figures 3.36 and 3.37.

Generally, most glow plug adaptors do not incorporate the heating function, and in most cases this is not an issue. Much engine development testing is done at operating temperatures, and hence the lack of cylinder heating is no issue. However, certain tests— obviously, cold start and running tests—can require the cylinder heating function, and recent technology developments have facilitated the development of a transducer and adaptor arrangement that includes the heating element. The compromise of this design is that the transducer measuring face has to be connected to the cylinder space via an indicating channel, due to the position of the heater element. This arrangement is therefore only suitable for engines with low or medium sound pressure levels.

Alternatively a specific design of measuring and heating glow plug has been developed that transfers the force from the cylinder pressure to the measuring element mechanically, via the glow plug tip. This prevents pipe oscillation but also allows remote positioning of the measuring element. Thus, the thermal shock effects are minimised, as the element is not exposed to the hot gases.

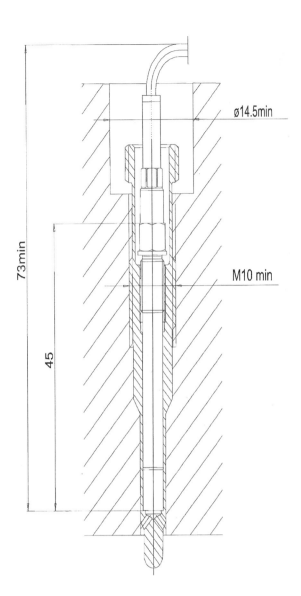

Figure 3.36 Glow plug adaptor and transducer.
(Source: AVL.)

With respect to the use of glow plug adaptors, there are some important points to consider:

The design of the glow pug and transducer will be a compromise with respect to accuracy and durability. The closer the measuring face is to the combustion space, the better the accuracy and the lower the impact of interference factors (pipe oscillation). However, this position exposes the transducer to a higher heat flow load, and for this reason, lifetime is reduced. It is important to consider this fact with respect to the measuring task. For example, if the task is just monitoring of the cylinder pressures in order to prevent excessive component loading (via maximum cylinder pressure values), then pipe oscillation can be accepted or filtered out. Hence, for this particular application, an indicating channel that

Figure 3.37 Glow plug sensor.
(Source: AVL.)

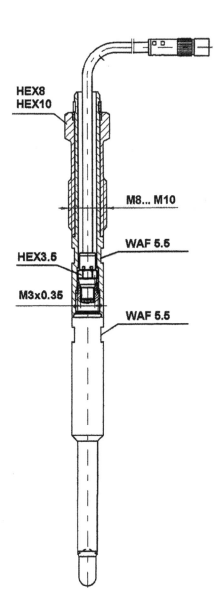

removes the transducer face from excessive heat flow is advantageous—particularly as monitoring tasks are generally part of engine durability tests that may be run for hundreds of hours. In this case, long life of the transducer to prevent any downtime would be more advantageous. It is clear that the chosen design of the transducer and adaptor used, must be appropriate according to the specific measuring application.

A critical dimension in the design of a glow plug adaptor is the concentric gap between the adaptor body and the glow plug bore. This is of great significance, as it dramatically affects the thermal load of the transducer. As the in-cylinder gas compression and expansion takes place, the thermodynamic and dynamic effects of the gas moving into and out of this clearance space increase the temperature around this area. The larger the gap, the greater the temperature increase and the greater the negative impact on the measured signal. For this reason, glow plug adaptors are normally made to bespoke dimensions, to fit the actual engine into which they will be installed. It is important, therefore, to have the glow plug bore measurements available when specifying the adaptor dimensions to the supplier. Note that the dimensions of the glow plug itself are not sufficient to ensure the optimisation and reduction of the annular clearance space. This is due the fact that the manufacturing tolerances of the actual glow plug are outside acceptable limits for a glow plug adaptor. If the actual dimensions of the installation bore are not available, then glow plug dimensions can be used with the acceptance of the risk of significantly reduced transducer life.

Always use the correct tools to assemble, repair, and re-install glow plug adaptors. In addition, observation of the correct mounting torque is essential for a successful and accurate measurement.

A glow plug–mounted sensor can be an ideal nonintrusive solution for diesel engine measurements, and the ongoing improvements in technology can provide a cost-effective installation that can fulfil nearly all measuring requirements. It is worth noting that manufacturers of transducers are currently working on glow plug measuring solutions that could be used as part of the standard sensors set for the electronic engine control system, providing feedback on the progress and quality of combustion to the engine control unit. Developments in this area are likely to increase the durability and reduce the price of glow plug as well as spark plug adaptors, assuming that similar (but possibly less accurate) transducer/adaptor units are fitted as standard equipment in production engines.

3.1.2.6 Transducer Selection and Applications

Introduction

Selecting the correct type of transducer for the measurement application and the installation environment is an essential task for the engineer prior to making the measurement, to ensure that the effects of the above-mentioned factors are minimised such that the transducer performance can be maximised with respect to measurement quality and reliability. The final choice will be a compromise among factors and boundary conditions that include:

- Installation space
- Installation location
- Measurement task, target, and focus
- Access pathways into the cylinder
- Cost
- Time
- Permanency of the installation

Taking these factors into account, the final choice of transducer and adaptor method can be chosen, and the required modifications executed, to allow access to the combustion space.

Requirements of the Application

Of course, the actual measuring task will have a significant effect on the choice of transducer. There is a wide variety of available types of transducer, and several of the available types could be used for a given application—although there will never be a perfect transducer that fulfils all required parameters for the task. In general terms, the transducer should be:

- As small as possible
- Easy to install
- High-sensitivity, good signal/noise ratio
- Resistant to the effects of the environment and application
- Durable in the application, with acceptable lifetime
- Once installed, no effect on engine operation

From the many different types of transducer, a suitable compromise can be selected. Generally, transducers can be categorised in the following ways:

- Threaded or unthreaded installation
- Size of body and mounting system
- Cooled or uncooled

Each category has its advantages and disadvantages, and these should be taken into account with respect to the actual measurement and evaluation. Figure 3.38 shows the relationship between the measuring task, the transducer choice, and the installation method.

Basic information about the operating conditions of the transducer and the measurement task requirement must be established in order to fully understand the requirements. The engine type itself, the focus of the measurement task, and the installation will all assist the engineer in choosing the correct type of transducer. Items to be considered are:

Pressure measurement range in normal operation for the measurement; also, over-pressure range, for a safety margin

Measurement task—factors that could reduce the life of the transducer (e.g., heat flows, abnormal combustion conditions)

Measurement target—combustion chamber, low or high pressure, inlet or exhaust system, low pressure measurements

Engine type—external factors that could affect the measurement quality; for example, high vibrations

Required level of accuracy—indirect or direct evaluation of the pressure curve

This last point is particularly important, as it can have a significant cost impact. Simple, direct evaluation of the pressure curve requires a transducer that is more robust but less accurate. These are cheaper to produce. A typical application is engine monitoring, where

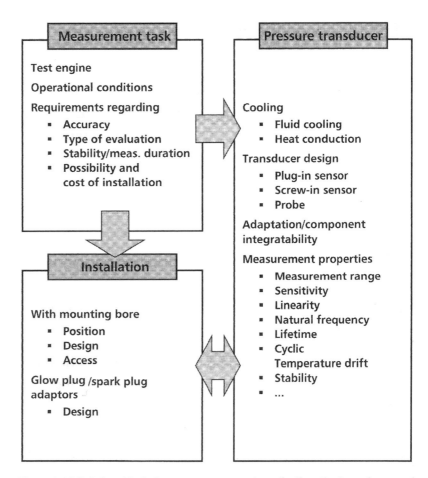

Figure 3.38 Relationship between measurement application, the transducer, and the installation site.
(Source: AVL.)

the curve is measured and evaluated simply for engine protection during unattended running on an engine test bed. Simple analysis of the curve will be executed in real time by the measuring system, to warn of dangerous operating conditions (e.g., detonation or excessive cylinder pressure).

An opposite and more demanding application is when detailed thermodynamic analysis will be carried out on the measured curves (either online during the measurement or offline in post-processing). In this case, accuracy, in absolute terms, is essential. Stability, accuracy, repeatability, and linearity are all important, and for this application, the largest, most stable measuring element is needed (often water-cooled transducers are specified). Of course, transducers incorporating these attributes are available, but they tend to be quite large. If they are to be fitted into a production-type engine, the challenge will be to fit them within the constraints of the engine design without compromising the combustion process, or the cooling and lubricating passages. Note, though, that the performance of such transducers is also favourable in the application, due to the excellent performance of these transducers with respect to long- and short-term drifting effects (i.e., sensitivity shift).

Size is an important factor. The ideal transducer would be very small with high sensitivity, but in general, size and sensitivity have a direct correlation. However, in certain applications, where high speed, vibration, or frequencies are encountered, then a small transducer with a high natural frequency (due to small size and mass) will be a favourable solution.

Categories of Transducers

As mentioned, transducers fall into basic categories depending on their main attributes and properties. In general they can be grouped according to:

- Cooled or uncooled
- Mounting type
- Size and sensitivity
- Defined characteristics (as per data sheet)

Water-cooled transducers are the traditional choice for maximum performance, stability, and accuracy. However, they are normally larger in size and require the infrastructure of a suitable fluid-cooling system. In addition, they can be susceptible to noise on the measured signal, due to the cooling water (cross-talk effects and resonance due to mass of cooling water). However, they are still a good choice for detailed analysis of the pressure curve, as mentioned above; they have superior performance with respect to drift and sensitivity shift.

Water-cooled transducers are often used in single-cylinder engine applications, where high accuracy is needed for combustion development work. Generally, in these engines, installation access is not a problem, as the cylinder head is designed for the installation of measuring sensors and probes. Hence, provision is generally made for the installation of this equipment. Simple screw-in, threaded bodies are often used, as the larger physical size is less sensitive to the effect of external stresses from the mounting position.

For greater accuracy, plug-in type water-cooled transducers are available. This design allows de-coupling of the sensor body from the mounting surrounding. Thus, stresses applied to the measuring element from the engine structure and mounting torque are eliminated as a source of error. In general terms, water-cooled transducers rely on the installation of a measuring bore. Due to their larger size, accommodating this may not always be possible in the engine under test. Generally transducers of this type are available in thread sizes of 8 to 14mm.

Uncooled transducers are smaller in size and are available as threaded or plug-in types. These are easier to install in modern engines and require no cooling system, which saves cost and complexity at the engine test bed. They normally have a reduced sensitivity compared to water-cooled units; however, the reduced size, apart from easier installation, means a higher natural frequency. Therefore, for applications at high engine speed or frequencies of interest, they are the preferred choice. The most common sizes are 5 to 7mm. The latter is available in threaded or plug-in versions. In addition, probe-type units are available that can be adapted to the most demanding or difficult applications, where cooling and lubricating channels must be crossed in order to gain access to the combustion chamber. These probe-type units are flexible in this respect and are available in different lengths to optimise and reduce installation effort. If possible, though, mounting bores or sleeves should be employed, as these assist the process of disconnecting the sensor element in the transducer from environment-related stresses that can affect the accuracy of the transducer as a whole. For this reason, direct mounting of transducers is to be avoided unless absolutely necessary. Note that uncooled transducers are generally employed in adaptor solutions for nonintrusive measurements where, due to their small size, they can be fitted into spark plug or glow plug adaptors. This reduces the installation effort by a considerable margin.

It is important to note that, although uncooled transducers have good accuracy and repeatability due to advanced crystal and manufacturing technology, they will still suffer with long- and short-term sensitivity shifts. This must be considered in the application.

The table in Figure 3.39 shows the main types of transducers and their properties with respect to selection.

Selection of Transducers for Common Applications

Once an understanding of the piezoelectric transducer and its properties has been gained, this can be considered with respect to the application, in order to select a suitable unit. The range of piezoelectric transducers for combustion measurement is quite wide. A logical

Pressure transducers for engine instrumentation

Type of cooling	Heat conduction (uncooled)					Fluid cooling	
Piezo material	$GaPO_4$					SiO_2	
Design	Sparkplugsensor M 10 / M12	Sensor M5 x 0.5	Probe Ø 4.3	Sensor Ø 6.2	Sensor M 8	Sensor Ø 9.9	Sensor M10 x 1 M14 x 1.24
Mounting principle		Direct (in cylinder head or components) or using an adaptor					
	Spark plug / glow plug					Requires extra space	
Metrological characteristics	as per specifications in the data sheets						

Figure 3.39 Categories of transducers and characteristics.
(Source: AVL.)

process has to be applied, to eliminate unsuitable types and narrow the decision margin to a limited number so that they can be compared and the compromises considered.

First and foremost the measuring task must be considered. What are the requirements, with respect to accuracy and stability, that are a function of this task? (E.g., measuring or monitoring, direct or indirect evaluation and calculation.) The next step is to consider the time and cost implications of modifying the engine to accept the installation of a transducer. In addition, the space constraints must be considered. From this information, one can decide whether access to the cylinder can or should be achieved with or without intervention. This basically means, should the engine be modified to accept the permanent installation of a transducer via a measuring bore or direct mounting installation, or should some kind of nonintrusive adaptor be used to gain access into the combustion chamber? Intrusive installation involves significantly more preparation work, in order to facilitate the installation of a transducer. In addition, depending on the size of the cylinder and the arrangement of the cylinder head casting, access in this way may be very difficult or impossible. However, many engine manufacturers design their engines from the prototype stage to include provision for access to the cylinder bores for use during the research, development, and calibration stages of the engine development process. There are few engines that cannot be modified in this way, and the wide range of transducer designs available means that, generally speaking, the limiting factor will always be time or cost. The benefits of a properly engineered adaptor installation are considerable. In many cases, the transducer measuring face can be placed in an appropriate position to achieve an objective measurement of the instantaneous, average cylinder pressure, representative of the in-cylinder conditions. Or it can be specifically placed to be sensitive to phenomena of interest. For example, positioning near the cylinder periphery helps in the measurement of knock modes and frequencies. As long as general rules regarding installation are followed (as discussed later), this installation method will produce the best measurement results.

The transducers available for this installation method include water-cooled and uncooled types. The actual choice depends upon user preference and space available. Water-cooled units are larger, and so it may not be possible to install a transducer of this type, especially where multivalve or small-bore engine applications are involved. In addition, a cooling system must be available. Modern uncooled transducers have performance comparable, in many applications, to cooled units. They are therefore an appropriate alternative and may be specified for the most demanding applications, in which ultimate accuracy is essential (e.g., for thermodynamic analysis).

When a nonintrusive installation is needed, the only possibility is small, uncooled transducers in combination with adaptors for access via the spark plug or glow plug. These provide a cost-effective solution but have some inherent limitations. The main problem is a result of the position of the installation of the measuring face of the transducer. This is fixed according to the adaptor type, and thus may not always be in the optimum location with respect to the requirements application. For example, when using centrally mounted spark plug adaptors for knock measurement, knock frequencies can be difficult to detect, because they occur as pressure waves around the periphery of the cylinder. Figure 3.40 shows a typical decision process that is followed when defining the transducer and installation type.

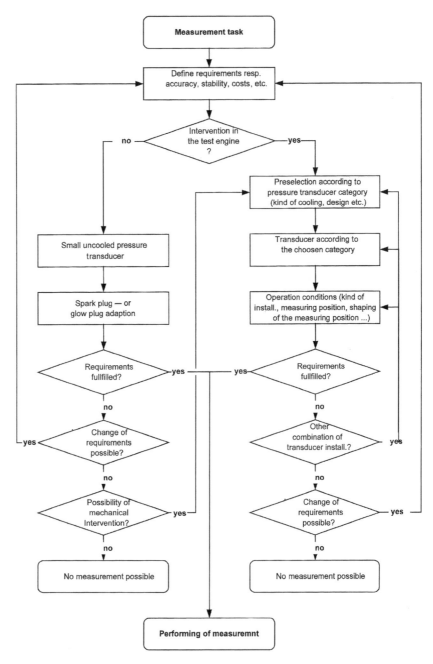

Figure 3.40 Pressure transducer selection flowchart.
(Source: AVL.)

The access method and transducer are chosen as a function of the measurement task, taking into account acceptable compromises that may affect the decision with respect to cost and time constraints. For the various typical measuring applications, there are a number of points to be considered:

Engine monitoring. In this application, absolute accuracy is less important; normally, a direct analysis of the curve is required to monitor the engine for abnormal operating conditions or faults. This is a common requirement where an expensive, hand-built prototype engine is operated in conjunction with an undeveloped ECU. The combustion measurement is used to monitor the cylinder pressure curve for maximum pressures (for component load limits) or, with a gasoline engine, for detonation/pre-ignition. Another typical, similar application is durability testing of engines, monitoring the cylinder pressures for the same reasons, to protect the engine.

As accuracy of the transducer is less critical, a lower cost transducer with lower specification crystal quality will be acceptable in this application, thus reducing costs. Nonintrusive adaptor mounting is also a preferred option if the measuring bore has not been pre-installed, as this also reduces costs. Uncooled transducers require no cooling system, and hence these are also commonly used. The most important factors for this application are that the transducer should be robust, durable, and resistant to the damage caused by abnormal combustion conditions. In addition, when mounting the sensor, thermal heat flow in the mounting position should be considered and, if possible, reduced (by careful mounting) to extend operating life.

Engine calibration. Similar to the above application, combustion measurement is used in this application to protect against the effects of abnormal combustion conditions that may occur while mapping the engine or tuning the calibration in the engine control unit. However, the transducers will also be used for thermodynamic analysis. 50% energy conversion and IMEP will be important results for the tuning and optimisation of an engine. Therefore, the transducers must be of suitable quality and durability to operate at the limit on normal combustion, for long periods, while returning reliable results for optimising parameters such as ignition timing (gasoline), injection timing (diesel), and valve timing (where this can be varied), as well as other engine parameters (exhaust gas recirculation). It is not desirable that the transducer should fail during a calibration exercise, as these can be very long—and consequently expensive—test runs that use many hours of test-bed time. In addition, when looking small variations in engine performance parameters, it is unfavourable to change a transducer mid-program, as this could introduce artificial differences in results due to variations in transducer calibration or performance, which could be perceived as errors or abnormal phenomena by the test engineer.

Typical sensor solutions are uncooled, as these have small mass and high natural frequency; thus, they are well suited for measuring high frequencies for knock and combustion noise evaluations. These are common optimisation targets, or boundary limits, when calibrating modern engines. These transducers can be mounted via measuring bores, but normally in this application, the engine is quite close to a production model, and measuring bores may not be available; thus, nonintrusive access is an ideal solution as long as any measuring compromises are well understood.

Engine knock (gasoline) or combustion noise (diesel). A very common requirement when developing modern engines is to understand these specific combustion phenomena. These may be useful as part of the engine calibration or monitoring processes, as mentioned. Or they may be the subject of a specific test to understand and develop the combustion system with these phenomena in mind. In both cases, the preferred transducers must have a high natural frequency in order to be able to measure the high frequencies for the evaluation. In addition, the transducers must be durable enough to withstand continuous operation at the limit of normal combustion. Generally, uncooled transducers with intrusive or nonintrusive adaptors will be the usual solution. Note the special requirements, during installation, to avoid interference or oscillation on the pressure curve caused by a "pipe" length in front of the measuring face. This is particularly important to avoid for combustion noise analysis, as the results will be completely inaccurate if additional high-frequency components are induced onto the measured curve due to pipe oscillations.

Cold-start measurements. For this application, understanding of the combustion events in the first rotation of the engine and the first few seconds or minutes of engine operation are important, and this has its own specific challenges for the transducer equipment. The sensitivity shift due to temperature change is an important factor; the operating temperature for the transducer may go from subzero to hundreds of degrees in a very short time frame. The transducer must exhibit an optimised behaviour with respect to this factor, and the calibration of the transducer must be carried out across the temperature range in order to understand the effects. In addition, the large change in engine temperature from a cold start, through warmup, to normal engine temperature, will cause the greatest amount of expansion in the materials surrounding the transducer installation. This will also have a considerable effect on the transducer's dynamic sensitivity. Use of a mounting bore in conjunction with a plug-type transducer will minimise this effect.

Resistance to electrical drift is also important, as drift compensation circuits in charge amplifiers cannot be applied at engine cranking speeds, due to the longer cycle time that could be affected by the action of the compensating circuit. Therefore, optimised long-term shift behaviour and resistance to charge leakage are important.

Uncooled transducers are ideal, as no cooling medium is required that could possibly freeze. In this application, nonintrusive adaptors are used, as engines are often production-near or -ready. If these are used, the appropriate heat range of the spark plug should be observed. For diesel applications, a glow plug adaptor that includes a heating element is necessary to maintain the engine starting and running at quasi-normal conditions.

Combustion and thermodynamic analysis. This is the standard measurement application; it involves direct and indirect analysis of the pressure curve. The particular challenge is with respect to calculation of important parameters derived from the pressure curve—for example, energy conversion and work done. Often, the engine will have multiple cylinders instrumented. Depending on the layout of the engine, different adaptor solutions may be used in different cylinders (depending upon access). If possible, though, where cylinder contribution is of interest, each instrumented cylinder should have the same transducer and mounting arrangement. This reduces the deviation between cylinders due to the instrumentation differences and allows an objective evaluation of each cylinder's work output when compared with one another.

For this application, cooled or uncooled transducers may be used. Both have their advantages and disadvantages, as discussed previously. However, the actual choice of unit will most likely depend on user preference or installation effort. The performance of uncooled transducers is so close to that of cooled transducers that, for general purposes, the additional effort of providing a cooling system is difficult to justify. If minimal intrusion is required, then clearly spark/glow plug adaptors will be attractive. An important point with respect to spark plug adaptors is that the heat range and electrode position are essential factors that must match the original, standard plug design. This is particularly important for direct-injection engines. In addition, mounting of the ignition coil wiring and transformers must be considered, as this can be a challenge where coil-on-plug ignition equipment is fitted, if tracking of the spark and misfiring are to be avoided. Note that there are designs of spark plug adaptor available that closely match the original plug and allow fitment of the original ignition equipment in the standard way.

Whichever transducer type is chosen, accuracy is important for the result calculations; therefore transducers with optimised behaviour with respect to intracycle and intercycle sensitive changes are important, as these have a direct effect on the derived results. In this respect, water-cooled units are popular, but uncooled units are a very close alternative, particularly with modern crystal technology and manufacturing processes. The largest possible transducer will be most appropriate (cooled or uncooled), as this will have the greatest sensitivity and hence, the best signal-to-noise ratio and behaviour.

Gas exchange and detailed energy balance. For a detailed analysis, the most accurate and sensitive transducer must be specified. One typical application is advanced or fundamental combustion development work, normally carried out single-cylinder engines. Generally in this application, access to the cylinder is via measuring bores, as these are always available in an engine of this type. They are required in order to achieve the needed level of accuracy and assist in de-coupling the transducer from disturbance factors (for example, heat and deformation). Access is generally good, and water-cooled transducers with high sensitivity are ideal. In addition, their excellent linearity and minimal dynamic sensitivity shift give good performance in both the low- and high-pressure parts of the engine cycle.

Friction measurements. This application has similar requirements to the above, although it could be applied in a multicylinder engine, which then could place some restriction on the mounting possibilities for the transducer. The main goal is to understand the gas exchange and high-pressure parts of the cycle, to establish work done and lost, respectively. In order to achieve this, the sensor must be sensitive, stable, accurate, and repeatable. Therefore, large sensors—water-cooled, if possible—should be used in conjunction with a measuring bore (if possible). Intra- or intercycle effects that have an impact on the dynamic sensitivity are important, and must be minimised to prevent distorted IMEP results on a fired engine. High sensitivity is important in order to capture the subtle pressure variations in the gas exchange process.

Generally the specification sheets from the manufacturer of the transducer will provide the information required to judge the best unit for a given application or task. Figure 3.41 shows the effect of transducer property and installation on the type of evaluation that will be carried out.

		Type of evaluation (measurement task)				
		Maximum, minimum, amplitude (cycle)	Qualitative curve before and after TDC	Frequency, amplitude of high-frequency oscillations	Integral values (energy conversion, mean)	Different values (heat release, dp/dα)
Effect of						
Pressure transducer property	Measurement range	W	W	W	W	W
	Lifetime (cycles)	W	W	W	W	W
	Sensitivity	C	W	C	C	C
	Linearity	Significant	W		Significant	Significant
	Natural frequency	W	W	W		
	Acceleration sensitivity	C	W	C	C	
	Shock resistance	W	W	W	W	W
	Temperature resistance (transducer)	W	W	W	W	W
	Change in sensitivity over temperature	C		C	C	C
	Cyclic temperature drift (heat flow pulse)	Significant	W		Substantial	Significant
	Zero-line gradient (load/heat flow change)	Significant	W			
	Zero-point deviation (load change)	Significant			Significant	
	IMEP stability (behaviour in continuous operation)	Significant			Significant	
Installation	Deformation	Significant			Significant	
	Indicating channel		W		Significant	
	Gas flow	Significant	W	W	Significant	

Key:

☐ No effect

W Warning! Effect only avoidable through careful choice of transducer, measuring position design, handling, etc.

C Calculable effect (e.g. 1% change in property means 1% change in pressure signal)

▓ Significant effect

■ Substantial effect

Figure 3.41 Effect of transducer installation and properties on the measured signal.

(Source: AVL.)

Chapter 4

The Measurement Chain: Additional and Alternative Transducers

4.1 Alternatives to Piezoelectric Sensors for Cylinder Pressure Sensing

4.1.1 Introduction

Combustion pressure sensing is most commonly implemented using sensors that employ the piezoelectric measurement principle in conjunction with appropriate hardware (such as a charge amplifier). Many users are familiar with this well-accepted technology. It should be noted, though, that with advances in technology and miniaturization of electronic components, alternative techniques and technology are available for use in transducers for the evaluation of the combustion phenomena via pressure sensing.

These lesser-used technologies can have certain key advantages over the piezoelectric measurement chain in some aspects of operation and service. In general terms, an ideal transducer for combustion pressure measurement should have the following features:

1. High sensitivity, accuracy, and linearity
2. Measurement of pressure in absolute terms
3. Small size
4. High natural frequency
5. Stable properties, irrespective of temperature, heat flow, and deformation
6. Minimal number of components and interfaces in the measurement chain

There is currently no single transducer or technology that can fulfill all these requirements for every application. The most widely used technology is piezoelectric, as described in the appropriate sections of this book. Other transducer and sensing technologies are available, however, and these should not be discounted, because they may be a desirable choice for certain applications. The decision about which transducer technology to use should be made on the basis of suitability for the application at hand, taking into account all the attributes of each available type of technology. The current alternatives to piezoelectric sensors (at the time of this writing) are described in the sections that follow.

4.1.2 Piezoresistive

Piezoresistive sensors are commonly used in combustion measurement applications for internal combustion engine development as sensors for measuring pressures in high-pressure fuel lines (in diesel engine development applications, for instance). They are less common for combustion pressure measurements, but they are nonetheless available for that application. In common with all pressure transducers, piezoresistive sensors use a force-summing device to convert the applied pressure into a stress or displacement that is proportional to the pressure applied. This stress or displacement is then applied to an electrical transducer element to generate the required signal.

The most prominent advantage of the piezoresistive sensor is the fact that this technology is capable of measuring static and dynamic pressures (DC and AC components of a signal); i.e., the sensor gives an absolute pressure reading at its output. Hence no reference pressure signal or zero-level correction of the measured signal is required in software. In addition, piezoresistive sensors can have their entire signal-conditioning electronic circuitry integrated. This fact simplifies the measurement chain considerably, with no additional signal conditioning hardware required between the sensor and the acquisition system.

The basic principle of operation of this sensor is analogous to a strain gauge. That is, when a conductor is strained, its length and thickness change (as predicted by Poisson's Ratio). Because electric current is forced to travel a longer path of smaller area, when the conductor is stretched by tension, the resistance of the conductor increases. The resistance change, compared to the original resistance, divided by the fractional change in length, is known as the *gauge factor*, or the K factor. A number of materials with different respective K factors are used in strain gauge applications, but a higher gauge factor means higher output for the same strain, or higher sensitivity relative to the stiffness and natural frequency of the structure.

4.1.3 Optical

Optical sensors were developed as an alternative to piezoelectric transducers for in-cylinder measurements. To clarify, this technology uses a fiber optic-based technique to detect the deflection (from the applied pressure) of a measuring diaphragm that is exposed to the combustion chamber. The light reflection from the flexing metal diaphragm is monitored via an electronic optical receiver (the light source is an LED—a *light-emitting diode*). The optical fiber system is completely enclosed within the transducer body. The output from the transducer is an electrical signal whose magnitude changes as a function of the change in applied pressure at the measuring diaphragm. This technique should not be confused with optical methods of accessing the combustion chamber for in-cylinder combustion diagnostics and visualization.

Most pressure transducer designs use a measuring diaphragm exposed to the target pressure. The pressure to be measured is applied at the diaphragm, and this pressure causes the diaphragm to deflect. If this deflection is applied as a force to a measuring element (i.e., to a piezocrystal in the case of a piezoelectric transducer) then, a signal can be produced that correlates to the applied pressure.

The optical sensor uses the principle of light reflection from the rear surface of a flexing metal diaphragm that is monitored by the optical system via optical fibers. The deflection of the diaphragm (which is a function of the applied pressure) is measured using an optical fiber strand bunch with transmitting and receiving elements. The transducer's measuring element responds to the pressure-induced diaphragm displacement, which in turn changes the optical signal transmitted from the sending fiber to the receiving fiber (the general arrangement is shown in Figure 4.1).

For a given diaphragm displacement caused by a pressure change, the sensor response, i.e., sensitivity and linearity, can be adjusted by appropriate selection of optical fiber-core diameters and layout. The position of the transmitting and receiving fibers relative to the diaphragm allows fine-tuning of the transducer's transfer function. In addition, the linearity can be optimised by adjusting the stiffness of the diaphragm. To guarantee durability, a proposed design by the manufacturer Optrand (www.optrand.com) uses a sculptured, hat-shaped Inconel diaphragm with a thickness that varies across its diameter.

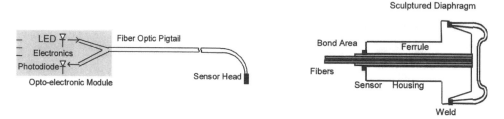

Figure 4.1 Optical pressure sensor, sensor construction, and sensor head.
(Source: Optrand.)

The diaphragm shape and material are selected to meet the requirements of high strength at combustion temperatures, with low creep and a fatigue life of well over a hundred million pressure cycles.

The measuring diaphragm should be as small as possible so that the transducer requires the minimal mounting space. At the time of this writing, the smallest sensor available has a diaphragm diameter of approximately 1.7mm. A small diameter creates a significant design challenge, though, because of the simultaneous requirement of large diaphragm deflection to achieve a high signal-to-noise ratio. Also, low stresses are required for a long diaphragm lifetime. For example, in a typical passenger car application, the sensor has to function reliably over hundreds of millions of pressure cycles, whereas in diesel truck engines the lifetime has to approach 2 billion cycles.

The advantage of the optical transducer technology is its compact size, which permits easy integration into existing engine installations. In addition, these transducers are highly immune to the effects of electromagnetic interference generated by the engine's ignition and electrical systems.

These benefits, combined with good durability and relatively low cost, mean that fiber optic sensors could be appropriate for use in automotive production engines for onboard diagnostics or combustion monitoring. Note that an embedded optical sensor does not require a separate access point into the engine, and the device can be integrated with another component (e.g., a spark plug) that can be conventionally installed. A typical installation of this sensor technology into engine components is shown in Figure 4.2.

Figure 4.2 Optical sensors integrated into engine combustion chamber parts
(Source: Optrand.)

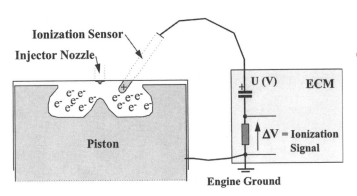

Figure 4.3 Electrical Field in the combustion chamber generated for ion current sensing
(Source: SAE.)

4.1.4 Ion Current

Ion current technology is well established for use in production engines, in addition to being used to some extent for engine research and development. This measurement technique provides information about certain aspects of the combustion process, even though it is not strictly a direct pressure measurement. It is worthy of mention, however, because it has been used to provide useful combustion-related information for engine management control loops, particularly for closed-loop control of engine knocking in gasoline engines, and for peak pressure monitoring in diesel engines.

The information gathered through ion current technology is often used in the development process to understand engine behavior and operational limits when one is mapping and calibrating an engine. However, in general, it is not current practice to equip an engine with combustion-measuring equipment during engine production (although that practice may be needed in the future, as we discuss throughout this book).

Ion current sensing uses a simple electronic circuit to apply a positive DC voltage inside the combustion chamber. Applying that voltage creates an electrical field, as shown in Figure 4.3.

This electrical field will attract the negative-charged species in the combustion gases. A small current is generated from the sensor to the electrical ground. The electrical ground comprises the piston and the combustion chamber walls. Figure 4.4 illustrates the basic principle of ion current sensing for combustion measurement.

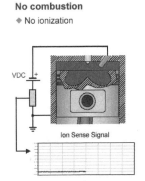
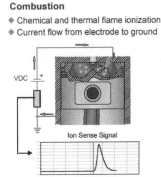

Figure 4.4 Basic principle of ion current sensing for combustion measurement
(Source: Delphi.)

95

Figure 4.5 A proposed glow plug design with integrated ion current sensing
(Source: Delphi.)

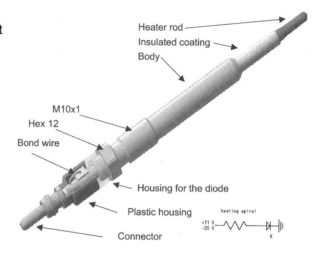

Glow plug sensors using this technology have been proposed and designed for use with diesel engines, as shown in Figure 4.5. As an alternative, a specific sensor can be installed directly in the engine.

In gasoline engine applications, ionization detection systems use a spark plug as a sensor to observe the in-cylinder combustion process. A bias voltage is applied between the spark plug center and the ground electrodes. Because the combustion flame is initiated at the spark plug gap and then gradually moves outward toward the periphery of the cylinder, the ionization signal can contain more detailed information about the in-cylinder combustion process with respect to flame kernel growth and trend than can an in-cylinder pressure signal.

When engine load is high, the ionization signal has a reliable and repeatable signature that can be used to locate the maximum pressure peak. A typical signal trace with respect to crank angle for a gasoline engine is shown in Figure 4.6.

Figure 4.6 Typical cylinder pressure curve and Ion signal versus crank angle
(Source: SAE.)

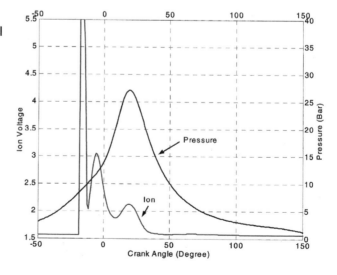

The first peak reflects initial flame kernel development right after the spark. When the flame front radiates away from the spark plug, the magnitude of the ionization signal reduces. As combustion occurs, the pressure in the cylinder increases rapidly. The combusted mixture around the spark plug gap is ionised again by the high temperature resulting from combustion, and that ionization generates the second peak. The third peak is caused by postflame activity. During experimentation, this peak shows good correlation with the maximum cylinder pressure peak position. In addition, research has established that, if there is no ionization signal after the spark event, the corresponding cylinder has misfired. Because of this knowledge, the ion current sensing system can easily be used to detect cylinder misfiring.

Many experimental systems and techniques have been investigated to extrapolate information from the ion current signal in a combustion engine, which could be used in engine control system loops. In addition to knock and misfire, partial burn and air-fuel ratio have shown correlation in experimental setups. There is clearly the potential to derive some interesting and valuable information for engine control system feedback. However, the most popular application of this technology in a production environment is still for the detection of knock, for spark timing adjustment, and for the detection of misfire for use in on-board diagnostics.

4.2 Other Transducer and Signals for Combustion Measurement Applications

4.2.1 Introduction

In addition to cylinder pressure measurements, it is of interest to measure and evaluate other phenomena during the the combustion measurement. These phenomena are generally dynamic signals for other systems whose activity directly relates to the combustion and in-cylinder processes, and that therefore require measurement in the same domain for comparison. The performance of these subsystems has a direct influence on the combustion process itself. These other subsystems therefore require high-speed, angle-based measurements to gain an understanding of their operation and effectiveness. The most commonly measured signals of this type are ignition-related signals for a gasoline engine (ignition timing signal, ion current signal) and injection-related signals for diesel engines (injector needle lift, dynamic fuel pressure).

An additional requirement may be measurement of valve motion. Such measurement is often required during cylinder head rig tests. A brief explanation of the technology commonly found in the measurement of these phenomena follows.

4.2.2 Ignition Signals

Ignition signals are measured channels of signals that relate to ignition and combustion events in a gasoline or spark ignition engine. The timing of the ignition of the fuel-air mixture is critical to achieving the greatest possible efficiency, in conjunction with the maximum power release available at the engine crankshaft. The ignition of the fuel-air mixture, the development of the flame front, and the combustion process must synchronise correctly with respect to the piston position and change in cylinder volume. For this reason, the ignition timing has to be optimised and adjusted with respect to a complex number of variables. That is particularly true of modern engine control systems that have considerable flexibility in this respect. In the simplest terms, the ignition timing must be adjusted with respect to engine speed and load, as a minimum. Therefore it is often necessary to measure the ignition firing point with respect to crank angle. Another area of interest, assuming a standard Kettering ignition system, is the coil *switch on* or *dwell time*. This is generally adjusted with respect to engine speed, because the coil charge time is constant with respect to time, and hence varies with respect to engine speed. It is therefore important to ensure that the ignition coil control system switches on and charges the coil for sufficient time to allow the buildup of energy, prior to the release of that energy at the required ignition firing angle.

The actual signal interface to the combustion measuring system depends on the ignition system type. Modern engines generally have a single ignition coil, or one transformer per cylinder, and these are switched via a low-level on-off pulse. This pulse can be recorded on the combustion-measuring device simply by connecting to this voltage signal via an appropriate amplifier. No transducer is necessary. In the past, less sophisticated systems used high-tension voltage distribution to each cylinder via high-voltage cables and mechanical, high voltage distributers (conventional ignition). These systems are not employed with modern engines. Transducers were available for earlier applications, and they may still be needed in some modern systems.

Such a transducer is an inductive clamp. These are widely available, and they are commonly used for ignition timing and tuning in workshop applications in conjunction with strobe lamps. These are typically directed at a mark on a rotating engine component to establish the basic ignition timing. Similar clamp-type units can be employed to measure the spark signal for combustion-measuring applications to provide the appropriate signal conditioning. Once the clamp is fastened around the high-voltage cable, it inductively couples a small electrical coil inside the clamp to the cable. The pulse in the high voltage cable induces a corresponding pulse in the coil inside the clamp. That secondary pulse is conditioned and amplified as the target signal. Due to the magnitude and short duration of this pulse, extensive conditioning is required (amplification, pulse shaping, and threshold determination). The output signal from a typical amplifier is therefore suitable only for spark firing determination (i.e., for determining ignition angle). None of the signal dynamic components can be captured with this measurement equipment because those components are filtered out during signal conditioning. This is less important because the

Figure 4.7 Inductive high-tension ignition clamp probe.

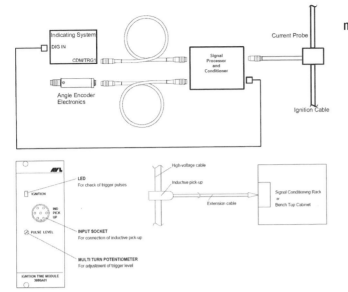

Figure 4.8 Ignition angle measurement equipment for combustion analysis
(Source: AVL.)

acquisition frequency of the angle-based measurement, even at high resolution (i.e., at a 1/10 degree crank angle) is insufficient to capture the high-frequency components of the signal that would show spark duration and energy dissipation in the ignition electrical system. Generally, though, for the application, the main target is ignition timing (spark firing) for correlation with combustion process events. A typical inductive probe is shown in Figure 4.7.

An inductive signal-conditioning measurement chain is shown in the schematic diagram in Figure 4.8. This includes clamp, cable, and amplifier unit. The amplifier responds to the ignition firing with a high, active, fixed-length pulse of suitable voltage for measurement by the acquisition system. The leading edge of the pulse is the reference point, and the trigger threshold from the signal is adjustable from 10 to 50%.

It is useful to note that the requirement for measuring ignition angle in this way is less common in an engine development environment. Most engine control systems (ECUs) generate ignition angle values during measurement, and these are generally used for comparison with combustion measurement results. In addition, most ignition systems no longer use high-voltage cabling that would require the use of a transducer of this type. Any electrical signal interfacing for data acquisition would be carried out on the low-level switching side.

4.2.3 Line Pressure

A very common requirement for diesel engines is an understanding of the dynamic behavior of the fuel system supply components that inject the fuel directly into the cylinder. The behavior of these systems is critical to efficient engine operation. Dynamic effects in these systems can have a negative impact on the process of introducing fuel into the cylinder. In modern, high-speed diesel engines, the time available for this process is limited because of higher engine speeds. In addition, the effectiveness of the fuel introduction procedure

Figure 4.9 Solder-on adaptor for line pressure measurements
(Source: AVL.)

is crucial, particularly with respect to engine cycle efficiency, because optimum engine performance requires that the compression-ignition combustion take place in the minimum time possible so as to achieve quasi-constant volume combustion where possible.

With this goal in mind, fuel line pressure is an important measured variable. It is common during research and development to instrument the fuel injector line with a high pressure transducer. Previous generation technology accessed the fuel pressure between the cylinder head-mounted injector and the injector pump (normally mounted separately on the engine, and driven by an auxiliary shaft at half engine speed). The line pressure and transducer mounting can be accessed via a saddle adaptor soldered onto the injector line. Once the saddle adaptor is mounted, a hole is drilled into the line to allow access to the hydraulic pressure at the transducer measuring diaphragm. A typical mounting adaptor is shown in Figure 4.9.

Application with this level of injector technology is straightforward, injection pressure is lower, and the line pressure is applied only during the fuel injection process. This procedure allows observation of fuel line pressure buildup and decay. In addition, this application permits measurement and evaluation of dynamic effects such as cavitations and hydraulic pressure waves.

Line pressure measurement requires different transducer technology from that used for combustion measurement. Thermal loading is less in this technology, but the hydraulic pressure applied to the transducer diaphragm is considerable. The physical size is tailored to maintain a high natural frequency for effective measurement of high frequencies, and to allow accommodation of the sensor via direct mounting on the pressure line.

The measurement technology most commonly utilized for line pressure measurement is piezoresistive—which measures the change of resistance of a semiconductor caused by mechanical stress, or a strain gauge—which measures the change in resistance of a conductive element caused by strain. Both techniques are applied to the measuring diaphragm, which deforms (as a force summing device) as hydraulic pressure is applied to it. These techniques use the deformation and consequent change in the diaphragm material to measure pressure change via integrated electronic circuitry. When one is using these sensors, one should note that they are passive rather than active sensors. That is, they need a power supply from the amplifier in order to produce a signal. The strain gauge requires a stabilised voltage supply. Piezoresistive technology requires a constant current supply. These facts should be noted when one is choosing and interfacing with signal conditioning equipment. In addition, the calibration certificate of the transducer generally states the power supply level that is required to achieve the correct behavior with respect to measurement accuracy, and this should be observed and applied during the measurements.

The latest generation engine technology provides new challenges for the measurement of diesel fuel line pressure. Common rail fuel injection systems are now widely employed, replacing the earlier standard pump and injector arrangement. This new technology affects the access point available for measuring the fuel pressure and thus the actual fuel pressure measured. The common rail system has a continuous high-pressure fuel supply, with electroactuated injectors. In addition, to meet current and forthcoming emission requirements, injection pressures are continually increasing, and injector operation strategies are becoming more complex with every new generation of injection system. The consequence of this is that the transducer must be able to withstand much higher pressures than previous systems had to deal with, pressure being continuously applied to the measuring diaphragm. Manufacturers have developed units that can withstand pressures in excess of 2000bar (at the time of this writing), and that are appropriate for use in these newer applications. It is also worthwhile to note that gasoline engine fuel injection technology has evolved. With the introduction of fuel injection directly into the combustion chamber (as opposed to the older system that used port injection), there is a requirement to measure fuel injection pressure (although the injection actual pressures are currently lower for gasoline direct injected engines when compared to diesel common-rail systems). The measuring chain components required for combustion chamber injection systems are most commonly either piezoresistive or strain gauge transducers, connected to an appropriate amplifier-signal conditioner.

4.2.4 Needle Lift

Needle lift is an important parameter that defines the point in time or in relation to crank angle at which injection of fuel into the cylinder occurs. This is a critical piece of information, particularly for diesel engines because, in conjunction with other combustion-related results, the ignition delay property of the fuel can be established (the Cetane Index). In addition, needle lift and line pressures (measured at multiple points on the fuel system hydraulics) establish the rate of injection and the dynamic behavior of fuel in the high-pressure fuel system at operating speeds and conditions.

In order to establish needle lift, one must instrument the injector itself. Many production diesel engines with electronically controlled diesel pumps are fitted with instrumented injectors to establish injection timing. A typical example from the Bosch company is shown in Figure 4.10.

This application uses an inductive method for generating the signal. Note, though, that sensors of this type are generally produced with design targets of low cost and service durability. The result of those design targets is that accuracy is compromised. The signal from such sensors is therefore more appropriate for indication (that is, for indicating sensor displacement) than for accurate measurement applications, as needed for detailed analysis of engine combustion in relation to injection system behavior.

Where more accurate measurements are required, as is commonly the case for research and development, many injection equipment manufacturers have the ability to instrument their own equipment in the prototype, or engine development phase when supplying their equipment to

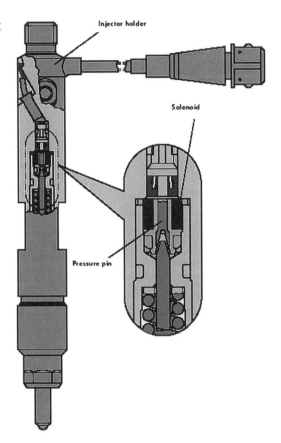

Figure 4.10 Production equipment needle lift sensor.

an engine manufacturer. This task requires detailed knowledge of the injector design, and the capability to engineer a sensing element into the injector body without affecting the operating characteristics or dynamics of the injector function. The need is to produce a signal of sufficient accuracy and repeatability, with favorable signal-to-noise ratio and acceptable durability for the application. This is a complex and time-consuming task that, in many cases, assuming that one does not produce high-pressure fuel injection system components in one's own company, is better left to specialist instrumentation companies.

Several technologies can be employed for needle lift displacement sensing. A carrier frequency signal is often used (see later section on Amplifiers for a description of this technology). The carrier frequency signal method can use tiny, wound, half-bridge sensors that can be installed in the smallest of fuel injectors to provide an appropriate signal.

A technology employed successfully by the Wolff company (www.wolffcontrols.com) is the Hall effect sensor. This sensor uses a magnetic or magnetised target (i.e., the injector needle), in conjunction with a Hall effect sensing element installed adjacently in the injector body. This sensing element generates a voltage signal proportional to the changing magnetic flux density caused by needle displacement. The output signal provided by this sensor is a differential voltage signal and this helps to improve the overall system noise rejection characteristics. In addition, the measuring element provides correction for temperature-related effects. Typical complete instrumented injectors are shown in Figure 4.11.

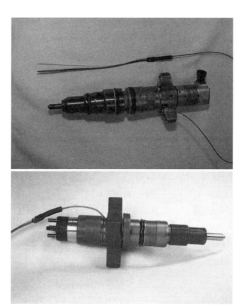

Figure 4.11 Instrumented Injectors
(Source: Wolff Controls.)

It is clear that an installation of this type, although requiring detailed engineering and instrumentation modifications, produces a favorable design that does not impede the actual installation of the injector in the engine. The only visible sign of the sensor element is the external cable.

A suitable amplifier must be used in conjunction with the instrumented injector to provide power to the measuring element and to condition the resulting signal for appropriate visualization or processing. A simple system overview of such a measuring element is shown in Figure 4.12.

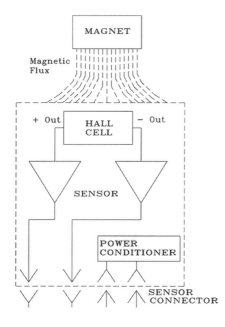

Figure 4.12 Hall effect sensing element interface connections
(Source: Wolff Controls.)

103

The signal derived by the unit, once in operation, contains static and dynamic components. When the moving target (needle) is at rest, the proximity of the magnet to the sensor creates a static voltage, depending on the strength of the magnet and the distance between the magnet and the sensor element. This static voltage provides a good indication of the condition of the magnet, and it should be observed and checked carefully over time while in use. In addition, the signal contains a nonmagnetic component, that is, an output voltage, when no magnetic flux is present. This offset voltage normally allows the sensor to measure movement in both directions away from a central rest position, depending on the relative position of the magnet and sensor after production.

The most important signal component is the dynamic component. This is the signal generated from movement between the sensing element and the magnet. The sensor is usually placed as close to the magnet as possible. This careful placement helps with linearization of the sensor output in order to make it directly proportional to magnet position. To achieve this desired effect, the magnet is usually made as powerful as possible.

Note that the target object generally moves a relatively small distance, so, while at rest, the magnet is usually quite close to the sensor, and the magnetic flux density is generally quite high compared to the amount of change of magnetic flux density as the object moves. For instance, the static component from a magnet at rest at a distance of 1mm from the sensor may generate a sensor output of 5 volts, whereas the dynamic component as the object moves may be only a few hundred millivolts. A typical output signal trace from the sensor is shown in Figure 4.13. Note the position of the static and nonmagnetic components.

Either of the signals can be used as a single-ended signal to be connected to a measuring device to display needle lift. A better option, though, is to use the two signals together to generate a differential signal. If that cannot be achieved via the data acquisition system

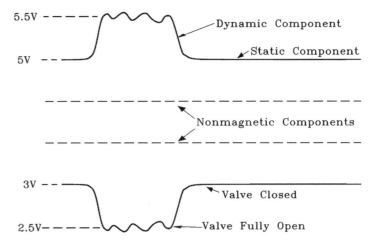

Figure 4.13 Signal output from needle lift sensor
(Source: Wolff Controls.)

(i.e., in a single-channel device), then an external amplifier can be used to convert the differential signal to a single-output with offset compensation. When one is processing the signal for digital conversion, the appropriate digital resolution is important. Because of the small amplitude of the dynamic component compared to the static component, a system with an appropriate dynamic range is essential to provide good quality digital conversion of the dynamic components, in conjunction with an appropriate range to acquire the entire signal range for subsequent analysis of the step function components for timing calculations.

Note that calibration of this signal in terms of absolute movement can be a challenge because the needle movement versus voltage transfer function can be difficult to establish accurately. For most needle lift applications, though, the most important aspect is the overall, relative movement and dynamic behavior of the needle during the injection process. In this case, establishing absolute movement is less important.

4.2.5 Valve Lift

It is often necessary to measure valve motion during the development of engine cylinder heads and in related component testing environments. In addition, it may be necessary to establish valve motion in a running engine, particularly in a research engine, where fundamental experimentation takes place. The situation is analogous to needle lift applications because of the similar operational environment, so similar technologies can be used to establish valve displacement, namely Hall effect, carrier frequency, and capacitive techniques. These sensor techniques, when used with appropriate signal conditioning, provide analogue signals that can be used to derive velocity and acceleration of reciprocating valve components. This is particularly important information for assessing the durability and lifetime of these components in the application. In addition, one must understand the dynamic behavior of the components at high engine speeds in order to optimise the timing and phasing of valve opening and closing events. This is necessary so as to be fully able to understand and optimise gas dynamic effects in the engine cylinder charging and gas exchange process.

Another interesting and commonly used technique is laser vibrometry, a noncontact measurement technique that provides a suitable analogue signal, derived from displacement of the target object (i.e., from an engine valve in the combustion chamber). The laser Doppler vibrometer is based on the principle of the detection of the Doppler shift of coherent laser light that is scattered from a small area of a test object. The object scatters or reflects light from the laser beam, and the Doppler frequency shift is used to measure the component of velocity that lies along the axis of the laser beam. Because laser light has a very high frequency, a direct demodulation of laser light is not possible. Therefore an optical interferometer is used to coherently merge the scattered light with a reference beam. A photodetector then measures the intensity of the mixed light, whose frequency is equal to the difference frequency between the reference and measurement beams. This signal can then be decoded to yield a displacement output (by processing the phase signal) and a velocity output (via FM demodulation). An overview of a laser system from the Polytec company (www.polytec.com) is shown in Figure 4.14.

Figure 4.14 Signal Processing of the Laser Vibrometry to produce velocity and displacement signals
(Source: Polytec.)

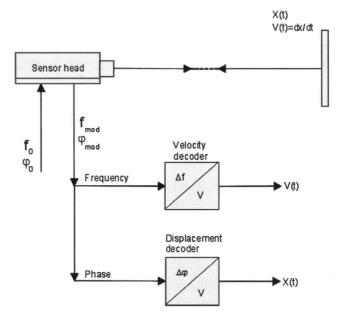

A typical system for cylinder head valve train analysis is shown in Figure 4.15. In this application, it is typical that one laser is directed onto the moving valve, and that the other, as a reference, is directed onto a static part of the cylinder head. This is a typical out-of-plane measurement application using a dual beam setup. This setup permits measurement of two points relative to each other. From this configuration, the valve displacement can easily be evaluated. Typical measured values, in a high performance engine that operates at up to 18,000rpm (at the time of this writing) are camshaft speeds of 9,000rpm, valve lifts of around 15mm, and maximum velocities of approximately 18 m/s.

Figure 4.15 Typical system setup for cylinder head component test of valve train
(Source: Polytec.)

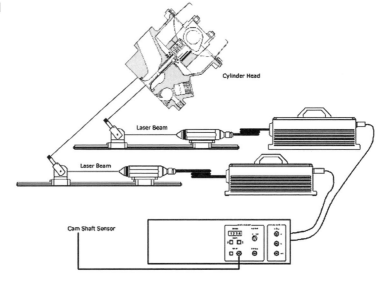

When the system is used in conjunction with a combustion measurement system, the analogue output from the laser vibrometry system, as a voltage, can be directly connected to and acquired by the analogue inputs of the combustion measurement system. This allows fast capture and storage of the displacement and velocity signals with respect to camshaft angle. Further derivation is possible to establish acceleration of the reciprocating components, and that information can be essential to establishing operation within fatigue limits and for lifetime prediction. In order to support such a measurement set-up, the combustion-measuring signal has to measure the rotational speed of the camshaft at high resolution, for example, at degree intervals. From this information, the camshaft angle-based data (velocity versus angle) set can be converted to time-based data (velocity versus time). From this data set, the acceleration can be derived simply from the numerical first derivative. In most combustion measurement equipment, this derivation can be performed in the user-specific calculation feature that most systems permit via the user interface.

4.2.6 Exhaust and Inlet Pressure

A common requirement for more detailed combustion analysis is to instrument the exhaust and intake system with suitable transducers so as to measure and evaluate the behavior of gas flow into and out of the engine. In general, this measurement provides an understanding of the dynamic behavior of the gas movement so as to be able to optimise it to achieve improved cylinder charging and charge motion, and in addition to be able to understand and promote efficient scavenging under operating conditions.

The information from these transducers is particularly important for measuring the gas-exchange or low-pressure phase of the combustion cycle. These measurements are used in conjunction with valve lift and high-pressure cylinder measurement to derive accurate energy release calculations, in addition to provide data for engine simulation models that calcuclate cylinder residual charge and trapping efficiencies.

This sophisticated evaluation process is becoming more and more important with the development of modern engine technologies to support current and future emission requirements, for which the required low levels of pollution are not possible via aftertreatment alone. Detailed understanding of the whole engine cycle is needed during testing to be able to understand the effect of new combustion technologies and strategies during the development and testing process. Typical positions for exhaust and inlet sensor mounting are shown schematically in Figure 4.16.

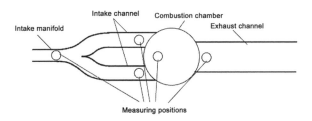

Figure 4.16 Measuring points for gas exchange
(Source: AVL.)

Figure 4.17 Cooling adaptor for exhaust mounting of a transducer
(Source: Kistler.)

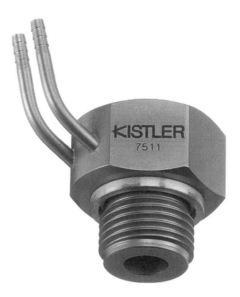

Instrumenting the inlet and exhaust manifold presents specific challenges to achieving accuracy and durability. For accurate measurement, the sensors should be located as close as possible to the engine valves. This placement prevents the effects of gas dynamics (i.e., travelling waves) from being superimposed on the measurement data.

In addition, the measuring face should, if possible, have no dead volume in front of it (as per a cylinder head installation) to prevent pipe oscillation. On the exhaust side, the transducer is subjected to extremely high gas temperatures. For this reason, the sensor must be located in a suitable cooling adaptor to ensure that the temperature of the measuring element does not exceed its working limits. Figure 4.17 shows a typical cooled adaptor.

Where possible, the sensor should be mounted in a straight section of manifold tube. Mounting the sensor in this manner prevents manifold pressure difference caused by the curve radius from affecting the results. If such mounting is not possible, mount the sensor at the midpoint of the curve radius to get a representative value.

The sensor must be highly sensitive because the gas fluctuations in the manifold are quite small when compared to in-cylinder data. High sensitivity is thus needed to achieve good signal-to-noise ratio of the measured curve and to facilitate high data quality after digital conversion and storage.

Another important factor to consider is acceleration effects. Because the inlet and exhaust manifolds are at relatively low mass (compared to the cylinder head mass), the transducer system itself has little damping from the structure, and it is thus subjected to extreme accelerations from engine vibrations. These accelerations cause unwanted noise on the measured curve, and it can be time-consuming to filter them out afterward. Therefore,

Figure 4.18 Cooled switching adapter for inlet pressure measurement
(Source: Kistler.)

acceleration-compensated sensors should be used where the vibration factor could be a problem (typically, in high-performance engines). In addition, mounting the sensor at an angle perpendicular to the direction of vibration accelerations improves measurement quality.

The establishment of absolute pressures in the inlet manifold is essential for accuracy of subsequent calculations and evaluations. The absolute pressure level must be established with an accuracy of 10mbar for a typical passenger car engine. Using piezoelectric sensors in this application is not preferable because they provide only a dynamic, relative pressure value. Modern piezoresistive sensors have a sufficiently high natural frequency to be able to measure the dynamics of the inlet gas pressure, but they also have the advantage of measuring absolute values; hence additional absolute sensors (required with piezoelectric sensors in the manifold) are not required. This makes piezoresistive technology preferable in this application.

Adapters for mounting the transducers can be used with inlet and exhaust measurements, and these can help to improve measurement data quality. A typical adaptor from the Kistler company is shown in Figure 4.18.

This adapter performs a number of useful functions. It acts as a damping adaptor and effectively isolates the sensor from structure-borne vibrations, improving the quality of the output signal. In addition, it provides cooling water channels to keep the sensor within operational temperature limits. That feature can be useful for inlet as well as for exhaust measurements. Modern engines with exhaust gas recirculation and inlet charge boosting can

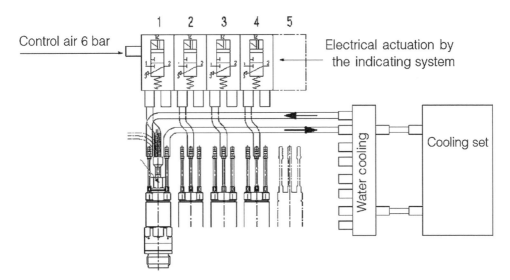

Figure 4.19 Cooling and switching adaptor layout
(Source: Kistler.)

operate with high charge temperatures, so water cooling protects the sensor and stabilises the signal, preventing drift and significant zero offset.

Though less commonly used, adaptors of this type often incorporate a switching function. This feature comprises a pneumatically operated switch that exposes the transducer measuring element, either to combustion chamber gases, or to ambient air. A typical system overview is shown in Figure 4.19.

The inclusion of this switching adaptor prevents the sensor from being exposed to hot gas for longer than the measurement time, thus extending unit life. The main reason for including this feature, though, is to facilitate the measurement of ambient pressure immediately before and after the primary measurement procedure. This technology permits accurate correction of the measured pressure curve based on measured values of ambient pressure, allowing temperature drift effects to be fully compensated in the subsequent calculations. Although this arrangement involves considerable expense and additional equipment, the extra cost can be justified by the improvement in data when accuracy is the main concern.

Chapter 5

The Measurement Chain: Measurement Hardware

The Measurement Chain: Measurement Hardware

5.1 Signal Conditioning

5.1.1 Introduction

The signal conditioning system is an essential part of the measurement chain. It forms the interface between the transducers and sensors, which measure the physical quantity of interest, and the measurement system that processes and stores the data from the measurement task. The importance of the signal conditioning system, as part of a measuring chain that produces high-quality data, cannot be underestimated.

The system conditioning amplifies the raw signal values from the sensors into a suitable, noise-free voltage signal, of appropriate amplitude, according to the measurement range of the physical value and input range of the measurement system. This amplification is an extremely important process. It must be done without adding any noise or frequency components to the signal, and the signal conditioner must also be capable of preventing and isolating unwanted noise from external sources. The amplifier acts as a firewall to protect the measurement system from noise and to transfer the signal in an efficient, unaltered, and predictable way.

This is a significant technical task when the environment of an engine test cell is considered. The other measurement devices and engine systems, plus electric machines and actuators commonly found there, are substantial electrical noise generators. In addition, the ground isolation characteristics in most engine test cell environments leave much to be desired. Ground noise–related issues are a very common source of disturbance on signal lines, and must be managed by the signal conditioner before reaching the measurement system.

Additional problems are caused by temperature effects. Because the signal conditioning is generally located close to the engine in the test cell, significant temperature changes can occur, depending on the test mode, and these changes have to be managed with minimal effect on the output signal to the measurement system. Moreover, the amplifier/signal conditioner system has to be intuitive to use and easy for the operator to set up; incorrect setup will result in the loss of data quality and integrity.

A typical signal conditioning/amplifier system for combustion measurement consists of a number of amplifier units, normally modular in construction, that mount in a rack enclosure. This enclosure is normally quite rugged and protects the more delicate electronic modules of the amplifier, which are normally plugged directly into the rack and supplied with appropriate, low-level DC (direct current) electrical supply from the rack itself. The rack also has a physical connector interface to allow connection with the measuring system. The connector interface for the transducer is generally on the front side of the amplifier. Adjacent to the connectors, for manual setting and parameterisation of the amplifier, switches and dials are available to the user. Often the rack is designed to industrial standards such that it will fit easily into the commercial electronic system cabinets often used in telecommunication, automation, and computing applications. Typically these racks are known as 19-inch (480-mm) racks. The total rack height is defined by the number of single-unit modules that it can accommodate. Units are known as HU (height units) at 1.75 inches distance. Typically, amplifier systems for combustion measurement are housed in 3-HU racks. More recently introduced systems are generally the more compact 1-HU systems. A typical rack of 3-HU design is shown in Figure 5.1.

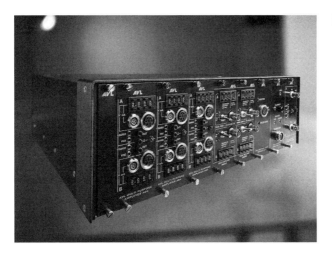

Figure 5.1 3-HU amplifier rack.
(Source: AVL.)

This modular concept allows amplifier modules to be moved around easily, replaced in the test environment, or with minimal intervention. The rack's power supply provides an isolated, clean power source for the amplifier's electronics. In addition, the rack is connected to equipotential and signal grounds to prevent ground loop interference effects on the signal line between the transducer, the engine, and the measurement system.

The latest generation of amplifier modules and racks consists of fully digital systems. As such, their size is reduced and their sensitivity to temperature change is less significant (or can be controlled/offset). Thus, they can be mounted in the cabinetry closer to the engine (i.e., the engine overhead boom for I/O interface), which reduces the length of the measurement cables (i.e., the cables between the transducer and amplifier, which are more susceptible to noise pickup). This improves the quality of the signal received by the measurement system via the signal cables (i.e., the cables between the rack and the measuring device). A typical example of a modern amplifier rack is shown in Figure 5.2.

Figure 5.2 Modern, compact digital amplifier system module: AVL MicroIFEM.
(Source: AVL.)

Digital amplifier racks normally have an intelligent interface to control and set the module parameters. This interface is often combined with sensor identification technology and plausibility checking of parameters to minimise the chance of operation with wrong setup values. An additional advantage is that the rack is mounted in the test cell: because it is not necessary to enter the cell in order to change the rack settings, there is an improvement in operator safety.

The value of good design and engineering of the amplifier rack and system modules should not be underestimated. A well-engineered rack with sensible and valuable design features to improve ease of operation and enhance the security of the measurement chain is an important benefit that should be considered when purchasing equipment from different suppliers. Harmonisation of the components increases efficiency of the workflow process.

5.1.2 Piezoelectric Signals—The Charge Amplifier
5.1.2.1 Basic Function and Operation

Probably the most commonly used amplifier for combustion measurement applications is the charge amplifier, also known as the charge-to-voltage converter. This is true because piezoelectric technology sensors are almost universally employed for combustion measurement in the engine cylinder—which in most cases is the main target—so piezoelectric sensors and charge amplifiers are nearly always present in the system measuring chain.

The charge amplifier has to condition the tiny charge signal into a voltage for processing via the measuring equipment. It has to do this without affecting the signal or introducing any unwanted filtering effects. In addition, it has to work in the most hostile of environments: an engine test cell.

As has been discussed in Chapter 1, a piezoelectric sensor delivers a small charge proportional to the change in stress in the measuring element when a force is applied to it via the sensor diaphragm. This charge signal is applied to the input of the amplifier, which then converts the signal into a voltage at an appropriate level for acquisition and processing by the measurement system.

In very simple terms, the charge amplifier consists of an inverting voltage amplifier with a very high open-loop gain and capacitive negative feedback circuit. It relies on a MOSFET (metal oxide semiconductor field effect transistor) or a JFET (junction field effect transistor) at its input to achieve the high insulation resistance required for minimal leakage currents. The schematic in Figure 5.3 shows the main working components inside a typical charge amplifier.

A charge amplifier comprises a high gain amplifier (A) and a negative feedback capacitor (C_r). When a charge is delivered from a piezoelectric pressure transducer (Q), there is a slight voltage increase at the input of the amplifier (A). This increase appears at the output substantially amplified and inverted. The negatively biased feedback capacitor (C_r) correspondingly taps charge from the input and therefore keeps the voltage rise very small at the amplifier input. At the output of the amplifier (A), the voltage (U_o) sets itself so that it picks up enough charge through the capacitor to allow the remaining input voltage to result

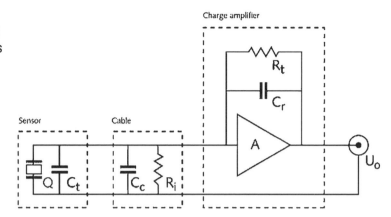

Figure 5.3
Main circuit components of a charge amplifier
(Source: Kistler.)

in exactly (U_o) when amplified by (A). As the gain factor of A is very large, the input voltage to the amplifier remains virtually zero. The charge output from the pressure transducer is not used to charge—i.e., to increase the voltage at the input capacitances—but is drawn off by the feedback capacitor.

With sufficiently high open-loop gain, the cable and sensor capacitance can be neglected; therefore, changes in the input capacitance—due, for example, to different cables with varying cable capacitance (C_c)—have virtually no effect on the measurement result. This leaves the output voltage dependent only on the input charge and the range capacitance. This can be expressed as

$U_o = -Q/C_r$

In summary, the amplifier acts as a charge integrator that compensates the sensor's electrical charge with a charge of equal magnitude and opposite polarity. This produces a voltage across the range capacitor that is proportional to the charge generated by the sensor element. That is, the main purpose of the charge amplifier is to convert the high impedance charge input into a usable output voltage.

5.1.2.2 Time Constant

The time constant property of the charge amplifier is an important criterion that defines the performance of the amplifier for quasi-static measurements. The time constant of a piezoelectric system is a measure of the time it takes for a given signal to decay, not the time it takes the system to respond to an input. That is, it defines the lower cut-off frequency of the system.

An additional resistance–capacitance (RC) network in the amplifier electronics defines the time constant, which is often adjustable for the user depending on the application (i.e., static or dynamic measurements). The time constant depends on the measuring range capacitor and the time constant resistor setting (adjustable to long, medium, or short), and can be expressed as

$TC = R_f \times C_f$

The lowest cut-off frequency can then be established from

$$\text{COF} = \frac{1}{2\pi \times \text{TC}}$$

Long operating mode (also known as DC mode) is used for quasi-static measurements and also during calibration of the amplifier with the transducer. Very long static measurements can be executed in this mode (thousands of seconds); note, though, that a too-low insulation resistance in combination with a zero error will cause excessive drift!

For cyclic combustion measurements, the medium or short setting (or drift compensated mode) should be used. In this mode there is generally less drift, though it should be noted that the amplifier behaves as a high-pass filter. Thus, when using this setting, measurements at low engine speeds may be attenuated or acquired with errors. As a general rule, the time constant must be 10 times longer than the duration of the measured cyclic signal. With these pre-conditions, the error due to attenuation of the signal will be less than 1%.

5.1.2.3 Drift and Drift Compensation

Drift, with respect to charge amplifiers, is a gradual change in the output signal that is not measured in the input signal. Drift of the output signal can be caused by a number of factors (e.g., temperature change, connecting cable properties), but a certain amount of electrical drift is always present because of the working principle of the charge amplifier electronics.

Figure 5.4 shows the effect of drift on the measured signal: the baseline of the measured curves without any compensation for drifting shows a gradual shift. This effect imposes a limit on the number of cycles that can be measured accurately during a single procedure. The drifting of the signal causes saturation of the input amplifier on the measurement device. In addition, to allow for some drift to occur, the input range of the measuring system cannot be fully optimised, leading to poor-quality digital conversion of the measured curve.

In order to offset this negative effect, most modern amplifiers are equipped with internal drift-compensation circuits. Generally, these work by applying a compensating current to

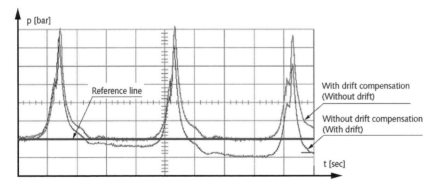

Figure 5.4 The effect of drift on the measured signal
(Source: Kistler.)

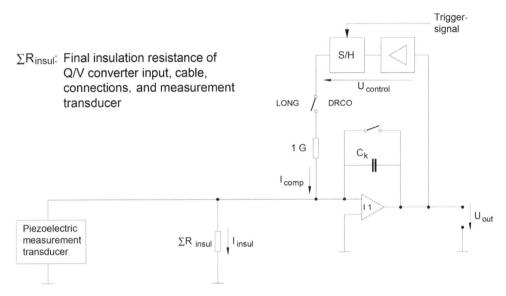

Figure 5.5 Drift compensation method
(Source: AVL.)

the charge amplifier input that is equivalent to the leakage current lost via the insulation resistances. One such system is shown in Figure 5.5.

This system consists of a control circuit (an input amplifier and sample-hold stage) that generates a control voltage ($U_{control}$) that feeds a compensating current (I_{comp}) to the input of the charge-voltage converter through a high-value resistor. This allows that the output voltage of the charge-voltage converter is kept at 0. If no input signal is applied to the charge amplifier, the sample-hold (S/H) stage is switched through and the control circuit ensures continuous drift compensation; that is, it clamps the charge-voltage converter output permanently at 0 level.

When an input signal is applied, a trigger signal is transmitted to the control circuit to identify the moment at which the output signal should be 0. These trigger signals are not applied to the sample-hold stage until the signal periods T are <1.2 sec—that is, when a minimum engine speed of about 100 rpm for a four-stroke engine has been achieved. This prevents the risk of the control circuit becoming unstable if the intervals between the trigger pulses are too long.

The trigger signals are generated internally from the input signal by means of a comparator and counter stages: the length of time between two pressure phenomena is determined, halved, and transmitted to the sample-hold stage as a trigger signal at $T/2$ of the next cycle.

Note that the specific technology for drift compensation depends on the amplifier manufacturer, but the most important fact to bear in mind is that the main focus should be to prevent electrical drift from occurring in the first place. This can be effected with "good housekeeping"—ensuring the cleanliness of all cables, connectors, and interfaces in the measuring chain.

5.1.2.4 Cabling and Interfaces to the Charge Amplifier

The magnitude of the signal produced by a pressure transducer is extremely low; therefore, good-quality cabling and interfacing between the amplifier input and the transducer output is essential to achieve the highest-quality measurements. In addition, sensible planning of measurement cable routing and cleanliness of the environment are important factors.

The main points to be considered are

1. Insulation resistance. Because of the magnitude and type of the input signal and the natural tendency of the charge amplifier to drift, insulation resistance must be maintained at high levels, generally greater than 10^{13} ohms for the complete charge signal chain.
2. Resistance to triboelectrical effects. Vibration and movement of coaxial cables produces electrical charge due to relative motion between the screen mesh and the insulator. This charge impinges on the measured charge, causing errors. For the charge signal, specific low-noise cable should be used that is manufactured with an additional, conductive layer to dissipate the unwanted charge from the triboelectricity.
3. Electrical screening. There are many sources of electrical interference in and around the engine test cell from the equipment, and from the engine itself. Cables must be routed sensibly and, if possible, kept away from all other electrical cable runs. Measurement cables and connectors must be kept clean and dry.
4. Durability. The cables and connectors must be fit for their purpose—capable of withstanding the high temperatures around the engine, and able to hold up under abrasion. They must be well screened with appropriate sealing and resistance to dirt. Connectors that allow multiple insertions/disconnections without wear or degradation of the signal are essential.
5. Cable length. Although cable length is insignificant with respect to the measured charge signal when a charge amplifier is used, excessively long cables can affect the upper cut-off frequency of the system. If measurement cables longer than 15 m are required, this effect must be considered. In addition, longer-than-necessary cabling causes problems with respect to storage of the cabling while in use; that is, preventing it from becoming contaminated may be difficult (excess cable is often left on the floor). In addition, the cable can act as a noise pick-up aerial. For these reasons, cable length should be properly planned and installed to ensure high measurement quality.

5.1.3 Analogue Signals

Often the measurement of additional phenomena around the engine subsystems during the combustion measurement can include the use of sensor technologies other than piezoelectric. Therefore, other signal conditioning units are needed in addition to the charge amplifiers. The types of signals most often found are analogue signals from the various sensor technologies available: piezo-resistive, hall-effect, strain-gauge, and the like. Normally, these sensors are passive; that is, the sensor requires a power supply in order to produce a signal. This is normally expected to be sourced from the amplifier module. The signal produced by the sensor must then be handled in the amplifier electronics to produce an appropriate voltage curve, of a suitable range, that maximises the use of the analogue-digital range on the measurement device. In addition, selectable filtering of the signal, and applying gain and offset to the signal, is considered to be an amplifier function. Note that two-channel modular units are very common and widely adopted (i.e., two channels per single module).

An important factor for any signal-conditioning unit for combustion measurement is bandwidth. Very-high-frequency components will be of interest in the measured data (e.g., dynamics of the hydraulic injection system for line pressure measurement on a diesel engine). The amplifier must be capable of processing these signals and producing an output of high integrity for processing by the measurement system.

Typical signals for analogue amplifiers to handle in combustion measurement are

- Voltage. Simple voltage signals from most sensors are not often in an appropriate range to be processed directly via the measuring device (in most cases this has a fixed input range of −10 to +10 V). Therefore, a simple voltage amplifier is needed to increase the magnitude of the signal and optimise it for analogue-to-aigital conversion. This should be done without filtering or distorting the signal in any way. In addition, the amplifier decouples the measuring system from the source. This is useful for electrical protection of the measuring system inputs from voltage spikes, and for preventing circulating ground currents from being superimposed on the measured signal.
- Current. Simple current signals and sensors are used where electrical interference could be an issue, or where long cable runs between the sensor and amplifier are needed. The standard operating range is 4–20 milliamperes (mA). This signal must be converted by the amplifier into a suitable voltage output. Gain, offset, and filtering are required.
- Bridge. Instrumentation bridge configurations are commonly found for measurement applications using strain gauges. These gauges are applied to components of the engine to measure stress during running conditions. There are various bridge configurations (Figure 5.6), but the amplifier is normally expected to provide the bridge with an appropriate voltage supply. The amplifier must them process the bridge signal, amplify it (applying gain and offset), and then produce a voltage signal to the measuring system. Strain gauges (Figure 5.7) are demanding applications, as they require a steady, precise voltage supply and a high gain factor from the amplifier.
- Hall effect sensors. Hall effect sensors are often seen in needle-lift applications, They require an appropriate excitation voltage in order to produce a signal. This signal then needs to be amplifyied before processing. Often the Hall effect sensor in needle-lift applications produces a complementary, differential signal, and this may require appropriate handling in the amplifier to produce the correct, ground-referenced signal for measurement or acquisition.
- Piezo resistive sensors. These sensors are common for pressure measurement applications (although not common for combustion pressure measurement). They require a stable excitation current in order to produce a voltage signal, which is then amplified for measurement and acquisition. Similar requirements for adjustable gain, offset, and filtering apply as with the other listed technologies.

The Measurement Chain: Measurement Hardware

Figure 5.6 Typical bridge configurations for a bridge amplifier
(Source: AVL.)

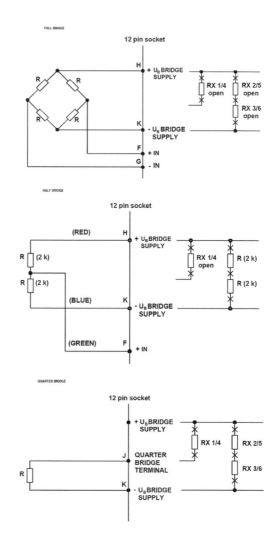

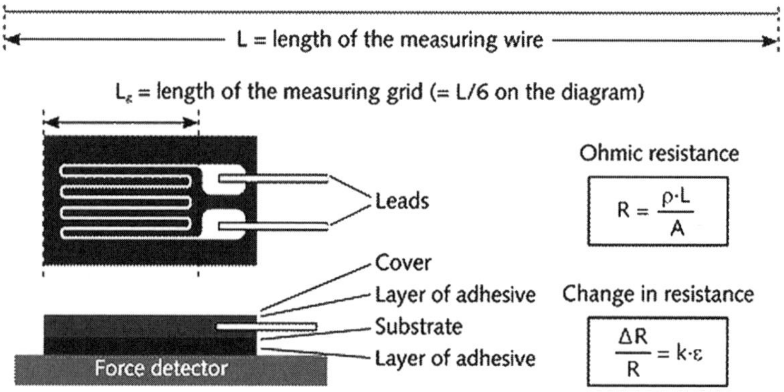

Figure 5.7 Typical strain gauge
(Source: Kistler.)

L = length of the measuring wire
L_g = length of the measuring grid (= L/6 on the diagram)

Leads
Cover
Layer of adhesive
Substrate
Layer of adhesive
Force detector

Ohmic resistance
$$R = \frac{\rho \cdot L}{A}$$

Change in resistance
$$\frac{\Delta R}{R} = k \cdot \varepsilon$$

Figure 5.8 Multipurpose amplifier for combustion measurement applications
(Source: AVL.)

Depending on the manufacturer of the equipment, analogue amplifier modules may be separate and specific for the sensor technology (i.e., bridge amplifier, piezo-resistive amplifier), and combined multipurpose amplifiers are also available These units can provide a voltage or current power supply to a sensor element, then process the signal via selectable amplification, gain, and filtering. AVL manufactures multipurpose amplifiers specifically for combustion measurement applications; an example is shown in Figure 5.8.

5.1.4 Other Amplifiers

There are a number of less common amplifier types occasionally seen in combustion measuring applications. These have been briefly mentioned in other sections of this book, but it is useful to reiterate them here.

5.1.4.1 Ignition Timing Amplifier

For gasoline engine development and calibration, ignition timing information is almost essential knowledge in a research and development environment. It is normally possible to gain ignition-related parameters (e.g., actual spark timing, offset from desired spark angle, coil charge time) directly from the engine control unit (ECU). However, in certain cases, it maybe useful to measure the spark timing alongside the combustion data, against crank angle. With older-technology engines, the ignition system often incorporated high-tension cabling, and it is possible to access the spark signal through a simple inductive clamp. This is a very common technique for diagnostic tools and equipment in the automotive aftermarket.

AVL produces an ignition-timing amplifier for use in conjunction with inductive clamps for measuring ignition angle as a crank-related channel. This is discussed in more detail in Section 4.2.2 of this book.

In this section the main focus is the amplifier. If it is possible to access the high-tension leads with an inductive clamp, it is connected to the amplifier's front side. The amplifier is available in a standard form factor to fit into a 19-inch rack with dimensions the same

The Measurement Chain: Measurement Hardware

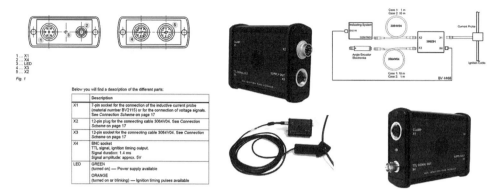

Figure 5.9 Ignition-timing amplifier
(Source: AVL.)

as other amplifiers of the same generation (10 partitions wide). An overview of a typical ignition amplifier is shown in Figure 5.9.

This amplifier has two main functions:

1. The ignition signal is a complex waveform with a spike of short duration that is developed by firing of the ignition spark. Even with a high crank degree resolution, and high engine speed, the acquisition frequency with respect to this signal is too low (as a crank-based channel) to prevent aliasing of the signal and loss of the true edge that indicates spark-plug firing. The ignition-timing amplifier receives the signal from the inductive clamp and modifies it to a clean, square wave pulse of 3 millisecond (ms) duration, approximately 3.5 V amplitude. The leading edge is the indicator for spark timing and can be easily detected and evaluated with a combustion measurement system. The trailing edge has no significance for the measurement, so the unit indicates only spark timing (not duration or duty factor of the ignition system).

2. Raw signals from any part of the ignition system's primary or secondary circuits have high-voltage, short-duration, transient pulses. If these were directly connected to the combustion measuring system, even via voltage amplifiers, the input stages of these devices would be damaged.

The unit has some adjustment to define the switching threshold for generation of the square wave pulse, because the ignition timing in a research and development environment is not generally derived directly from the high-tension ignition system, and because most vehicles use a coil-on-plug system where there is no access to the high-tension system. The ignition amplifier is rarely seen in use at engine test cells.

5.1.4.2 Carrier-Frequency Amplifier

The carrier amplifier is less commonly seen at the engine test cell. However, there are certain applications where it is suitable, particularly for measurement of rapidly changing mechanical events in the range of 0–20 kilohertz (kHz). A carrier amplifier of the type shown in Figure 5.10 (specifically designed for use at the engine test cell) is used in

Fig 5.10 Carrier amplifier for use in engine test environment
(Source: AVL.)

conjunction with measurement transducers in a half-bridge circuit. Inductive, capacitive, and resistive measurement transducers with an impedance of 1.2 Ω to 1.2 kΩ at 100 kHz can be used. The main application area for this model is for use with inductive measurement transducers in a half-bridge circuit with >2 × 50 Ω. Typical applications in engine instrumentation are injector-nozzle-needle and valve-lift measurements.

This amplifier is similar in form to other amplifiers from AVL of the same generation (i.e., 19-inch 3-HU rack mount, 10 partitions). Note, though, that it is a single-channel amplifier (as is the AVL ignition amplifier).

The basic principle of operation is as follows. The measurement of transducer supply voltage (carrier frequency 100 kHz) is generated in a Wien-Robinson oscillator. For amplitude regulation, the oscillator voltage is rectified and compared with a reference voltage. At a resistance of at least 50 Ω, the oscillator supplies a voltage of approximately 14 Vpp. The modulated carrier frequency is true phase rectified in a demodulator circuit. A two-stage active low pass filter suppresses the remaining part of the carrier frequency and provides a smoothed output signal. An overview of the internal signal processing is shown in Figure 5.11.

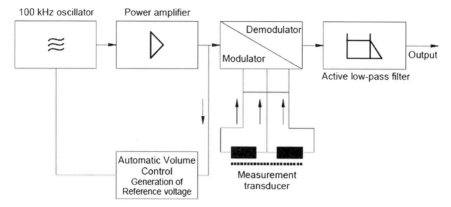

Figure 5.11 Schematic diagram of carrier amplifier
(Source: AVL.)

In use, the carrier amplifier is a sensitive piece of electronic equipment. It needs to be operated in a stable temperature environment. In addition, the setup of the system includes balancing adjustment of the signal and calibration before use. These are not particularly user-friendly procedures, and they require a full understanding of the working principle of the amplifier by the user. If this is not the case, the likelihood of a successful measurement is reduced. For this reason, the carrier amplifier is less common, and other technologies are more commonly used for the applications (Hall sensors or laser vibrometry).

5.1.5 Intelligent Amplifiers

5.1.5.1 Introduction

What is the definition of an intelligent amplifier? Previous amplifier technologies, as those discussed above, generally employ analogue electronics and signal processing to achieve the required signal conversion and amplification. As developments in digital electronics have increased digital signal-processing performance and calculation speed, in addition to reduced costs, these developments have been applied to combustion measurement signal-conditioning systems to produce a new breed of compact amplifiers, with advanced features that improve performance, allow easier setup with less chance of undetected errors, and optimise and harmonise the signal conditioning as a fully integrated part of the measuring chain for combustion measurement.

Because these units are generally based on digital technology, they provide a high degree of freedom with respect to the scaling factors and filter properties that can be applied to the measured and output signal. In addition, intelligent interfaces between the measurement system, amplifiers, and the user interface mean that the signal conditioning forms an integrated part of the measuring chain that can be parameterised and monitored by the user without having to have proximity to the system; this is important where the signal conditioning is located in the test cell close to the engine. These systems are compact and modular in design to facilitate a flexible, scalable architecture that allows the best possible diversity and flexibility of the available equipment in the test environment.

Figure 5.12 AVL MicroIFEM and Kistler SCP compact digital amplifier racks.

All these features, however, constitute just a first step; the latest generation of intelligent amplifiers can now include certain levels of simple signal evaluation. This second step means that the amplifiers themselves are now becoming a measurement system (to a limited extent), and for a user that needs only simple monitoring, this feature can save significant cost. Also, intelligent amplifiers that can recognise sensors and record operational data are now available. This facilitates a greater understanding of the lifetime of sensors and minimises the chance of incorrect parameterisation of a channel. Two examples of current generation digital signal conditioning systems are shown in Figure 5.12.

5.1.5.2 Sensor Recognition

One of the major advantages for intelligent signal conditioning is that it is possible to create an environment where sensors can be identified and monitored. Their information can be stored electronically and then accessed and used during the measurement process to assist with parameterisation and calibration of the channels. In addition, information regarding the operation of the sensor can be gathered and stored to help understand the sensor lifetime and to maintain correct calibration over the life of the sensor.

To fully capitalise on this feature, a completely harmonised environment must be in place. Sensor recognition is only part of that environment. There must also be an intelligent way of handling the sensor data; that is, a sensor database must be in place and accessible to all users of the equipment, not just to those responsible for calibration.

TEDS

An international standard, proposed by the Institute of Electrical and Electronics Engineers (IEEE), for so called smart sensors is currently being accepted within industry. This standard is known as Transducer Electronic Data Sheet (TEDS). The idea behind this concept is that, as sensors gradually develop smart features, the necessary protocol behind the communication should conform to a standard. The implication of this is that any smart TEDS sensor should be capable of being identified by any smart signal conditioning equipment, and the information on the sensor should be useable by any system.

The TEDS is stored electronically within a microchip that is so compact it can be accommodated in the sensor itself. Alternative locations are on the sensor cable, or within the sensor connector. A typical example of a TEDS-enabled connector is shown in Figure 5.13. The advantage of this flexibility is that new sensors can be purchased with

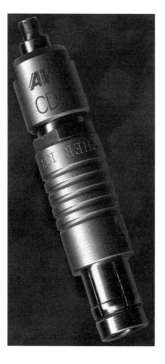

Figure 5.13 Connector with integrated TEDS microchip
(Source: AVL.)

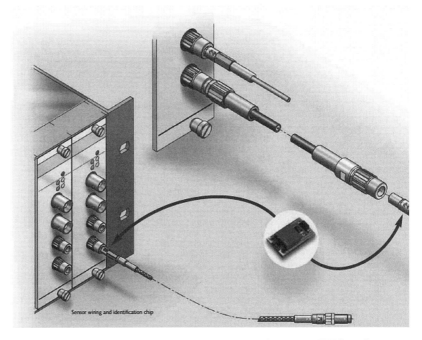

Figure 5.14 TEDS environment
(Source: Kistler.)

TEDS integrated, whereas existing sensors can be retrofitted with TEDS via replacement of the cable or connector.

When the sensor is connected to the signal-conditioning system (Figure 5.14), the data are read and available to the user via TEDS software or, in the case of combustion-measuring systems, the user interface of the measurement system (most manufacturers support this). The user can store the information and also write information onto the TEDS chip if, for example, a new calibration is carried out.

Data security management can also be implemented with the TEDS concept. That is, access user levels can be assigned to the TEDS data such that operators can read data only and calibrators can read and write data. Manufacturers could read/write at the highest level to store the information regarding the sensor type, initial calibration, and so on, which, if stored on a TEDS chip actually located on the sensor, will remain the same throughout its life. Such security organization can prevent accidental mismanagement or overwriting of important data. TEDS data can be stored and managed from a central database if this is required or available; doing so can assist with management of the calibration environment. The advantage of TEDS, though, is that this centralization is not strictly necessary: all the information the sensor needs for setting up the measurement system is stored on the TEDS chip, so if no connection to an upstream database is available, the measurement procedure is not impeded. An example of a TEDS data sheet is shown in Figure 5.15.

Transducer Electronic Data Sheet
TEDS

Basic TEDS	Manufacturer ID		
	Model Number	41	
	Serial Number	462992	
	Version Letter	53e	
Standard and Extended TEDS	Calibration Date	April 22, 2002	
	Temperature effect on span	0.0045	
	Temperature effect on offset	0.0045	
	Min Operating Temperature	-53	
	Max Operating Temperature	121	
	Response Time	0.0005	
	Min Electrical Output	-2	
	Max Electrical Output	+2	
	Sensitivity	1.998	
	Bridge Impedance	350	
	Excitation Nominal	10	
	Excitation Maximum	15	
	Excitation Minimum	3	
	Max Current Draw	30	
User Area	Sensor Location	23 right dyno	
	Calibration Due Date	April 21, 2003	
Templates	Special Calibration Data	12.3=0.175x+0.00563x	
	Wiring Code	Wiring code #15	

Figure 5.15 TEDS data sheet.

Contained in the standard is the definition for TEDS, which allows the use of very small memories via the use of templates. The small physical size of the memory device allows a TEDS microchip to be integrated in tiny, lightweight sensors. However, the low bit count available in physically small memory devices dictates that only essential data can be stored in a data array governed by fixed a template. The template defines the significance and units associated with the stored data and the mapping of the data in the chip's memory. Prior to

Figure 5.16 Standard templates used for TEDS
(Source: IEEE.)

IEEE standard templates

Type	Template ID	Name of Template
Transducer Type Templates	25	Accelerometer & Force
	26	Charge Amplifier (w/ attached accelerometer)
	43	Charge Amplifier (w/ attached force transducer)
	27	Microphone with built-in preamplifier
	28	Microphone Preamplfiers (w/ attached microphone)
	29	Microphones (capacitive)
	30	High-Level Voltage Output Sensors
	31	Current Loop Output Sensors
	32	Resistance Sensors
	33	Bridge Sensors
	34	AC Linear/Rotary Variable Differential Transformer (LVDT/RVDT) Sensors
	35	Strain Gage
	36	Thermocouple
	37	Resistance Temperature Detectors (RTDs)
	38	Thermistor
	39	Potentiometric Voltage Divider
Calibration Templates	40	Calibration Table
	41	Calibration Curve (Polynomial)
	42	Frequency Response Table

storing data and upon reading back the stored TEDS data, the template guides packing and unpacking the data, respectively. The standard templates defined for various sensor types are shown in Figure 5.16.

To access TEDS information, a specific connector must be available as an interface for the data and for the measured curve on the signal-conditioning amplifier. The connector shown in Figure 5.13, in addition to housing the TEDS chip, fulfills this function. The BNC connector normally used for a charge signal input on a charge amplifier (as an example) is supplied with a parallel input connector that supports the measured analogue signal plus the data connection interface to the TEDS chip. An example, from an AVL MicroIFEM, is shown in Figure 5.17. This unit clearly shows the two signal interfaces, one with signal only and one with signal and data. Other manufacturers of similar equipment use the same topology.

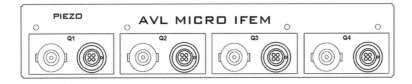

Figure 5.17 AVL amplifier with parallel inputs for signal only and for signal with data connection
(Source: AVL.)

Q1 to Q4	Connection for pressure transducers
Lefthand socket	BNC socket for pressure transducer with BNC connector
Righthand socket	4-pin **FISCHER** socket for pressure transducer with Sensor Data Connector (SDC)
4 LEDs	Status display of each amplifier channel
	GREEN: Regular operating state
	ORANGE: Acknowledgement of a control signal
	RED flashes: The amplifier channel is in saturation.

Figure 5.18
Concept of
data and signal
transmission
via SAW.

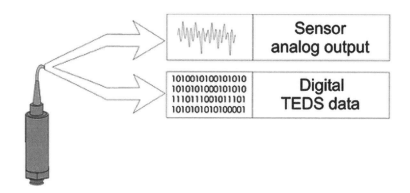

An alternative to TEDS for sensor recognition has been developed by the AVL company. This technology, called SID (sensor *id*entification), avoids the use of specific connectors for data transfer. The data chip is located on the sensor, and it holds only a very small amount of information, just the serial number. This information is read along the signal line with surface acoustic wave (SAW) technology. The concept is shown in Figure 5.18.

The advantage of this system is that integration of the chip into the sensor package has low cost, because there are no specific modifications to the wiring or sensor housing. It is included as standard on all AVL sensors recently produced. When the sensor is connected to the amplifier with SAW active, the acoustic wave signal is generated and received by the chip, which then transmits the serial number back to the amplifier. Once this information is received, the transducer is referenced to a database where all calibration and operational data are stored and easily available via the user interface of the AVL combustion-measuring equipment (AVL Indicom) or via the stand-alone software supplied with the signal-conditioning racks. An integrated sensor database is essential to this concept, because there is limited information recorded on the sensor chip (just the serial number). It is essential that the user have access to a database; otherwise the system has no information about the sensor. Thus, each measurement system has its own sensor database installed that is a replication of a master database maintained by a central calibration environment. Every time the measurement system connects to the network, the sensor database is automatically synchronised with the master. The theory is that all measurement systems have the same, latest information from the master database but it is still possible to work remotely if necessary.

An important feature available in most signal-conditioning systems for combustion measurement is the ability to record operating data of the sensor, such as hours or cycles, and to store this information. This information is particularly useful: tracking of operating hours of individual transducers is a difficult task, yet very important for planning regular checking and recalibration, as well as for assessing the suitability of a transducer for a test program, or deciding if the transducer can be used for accurate measuring or just monitoring.

This tracking could be done simply by monitoring the time that the sensor is connected to the amplifier and then recording the information on an ongoing basis. This method would be too simplistic and misleading, though. Often, an engine may be in a test cell, waiting to start a test in a paused condition, perhaps waiting for measuring devices or conditioning system to reach an operating or test condition. During this time the sensor would record operating time, because it is connected to the amplifier that is powered up, even though the engine is not running and the sensor is not experiencing the cyclic forces and stresses that cause wear.

A more sophisticated approach is to monitor when the engine is running. The amplifier can do this by evaluating peaks on the incoming signal trace that show that the engine is operating, even if measurements are not being taken. In this way, the actual operating time can be recorded as operating hours or engine cycles. Thus, the data will reflect the true operation of the sensor for lifetime monitoring, as it is only when pressure is applied to the diaphragm that the sensor is effectively working and experiencing wear. With this information, regular calibrations can be effected after realistic operating times, ensuring that the sensor is monitored closely for changes in its properties that could indicate failure. This is most important because failure halfway through a long test run or test program is far more expensive than the cost of replacing the sensor itself. In addition, data gathered over longer periods can be used statistically for intelligent planning of sensor life with respect to test programs and purchasing strategy. The only shortfall in the quality of this operational data is the fact that the actual test conditions are not known—only that the sensor is being subjected to cyclic pressures. Overpressure, temperature, and combustion knocking measurements are all engine operating conditions that can seriously affect the predicted lifetime of the sensor. If a test mode contains a significant number of these conditions (for example, engine calibration requires that the engine is regularly taken to borderline knock conditions for tuning of the ignition maps), then the sensor's lifetime should be expected to be reduced. This has to be considered with respect to the recorded operating hours.

For combustion-measuring application, sensor recognition is a valuable asset. The ability to identify the sensor specifically, combined with calibration data and operating hours, is a powerful concept that protects the measurement procedure from these risks:

- Incorrect parameterisation
- Use of the wrong sensor
- Use of a sensor that is out of calibration or near the end of its life

This ability can have a positive effect on the efficiency of the whole test environment. The interfacing of signal conditioning with sensors and measurement systems therefore becomes an important point to note. For the concept of sensor recognition and central management to work, the elements of the system must be fully harmonised; in simple terms, that means the sensors must communicate with the signal conditioning, which must in turn be able to talk to the measurement system and the database. This is an absolute prerequisite! To achieve this harmonisation in a risk-free way, all elements could be purchased from one supplier, but in practical terms this is rarely the case. Therefore, to maximise the benefits, make sure that all interfaces on equipment are defined and open. In addition, make sure that all current sensor stocks can accommodate sensor recognition in a retrofit manner without great expense. This will prevent the implementation of sensor data management from being a painful and expensive experience.

5.1.5.3 Extended Functions for Monitoring and Measurement

PMax Monitoring

The next step in development of intelligent amplifiers already present in the marketplace (at the time of writing) is to integrate some measurement or monitoring functionality, although only to a limited extent. The driving force for this requirement is that, for certain applications, a full measurement system is an expensive option that adds unnecessary complexity. An example is engine monitoring when durability testing an engine. Obviously, it is useful to protect the engine itself from overload, overheating, and the like during a test. This is normally effected via the usual measured channels of temperature and pressure on the test cell automation system. The problem is that cylinder overpressure will damage the engine within a few engine cycles. This is a critical factor for diesel engines: the engine can be destroyed before external temperature and pressure sensors register an abnormal condition to trigger an alarm. In this application, it is useful to monitor peak pressures in the cylinder. This would normally involve the use of a full combustion-measuring system with associated signal conditioning and encoder. A more appropriate alternative would be a device that can perform a simple evaluation of the pressure curve in real time, with online calculation of the peak value for display or output. Such a function can be integrated into a digital amplifier using the internal signal processor for measurement and evaluation. A unit such as this, apart from being much cheaper than a complete combustion-measuring system, is easier to integrate into the test environment to perform this relatively simple analysis. This concept is realised by several manufacturers of combustion-measuring equipment with intelligent amplifier concepts; examples are shown in Figure 5.19.

The amplifier, connected to the cylinder-mounted transducer, converts the charge signal into an analogue voltage output for connection to a measurement system if required. Typically, all standard amplifier functionality is supported (drift compensation, filtering, ground isolation, etc.). In addition, the unit evaluates the incoming curve and extracts the peak value as a cyclic result. This result can then be displayed on a small screen for viewing (if fitted), but more importantly, the result can be output through a digital-to-analogue converter as a voltage (for recording or monitoring). It can also be output as a digital bit should a defined event occur (e.g., exceeding a user-defined threshold) for fast reaction by the control/automation system to protect the engine. Simple peak-monitoring devices for engine protection are relatively new in the research and development environment for vehicle engines, because in this environment a combustion-measuring system normally exists. This simple measurement and analysis equipment is commonly found on large, ship, or stationary engines for continuous monitoring, performance evaluation (balancing), and protection of the engine. A typical hand-held system is shown in Fig 5.20.

Note that the requirement for monitoring peak pressure is quite simple: the algorithm to extract the peak value is simple and can be easily executed on a small signal processor. In addition, no angle encoder is needed: the crank-degree angle needs only approximation, which can be done by evaluating the time difference between peaks (this also gives an approximate cyclic engine speed result). Once the signal is correctly approximated relative to the crank degrees, it can have the appropriate zero-level correction (or pegging) algorithm applied for establishment of the peak pressure value in absolute terms. An alternative for

Figure 5.19 Peak-monitoring
intelligent amplifiers
(Source: Kistler and AVL.)

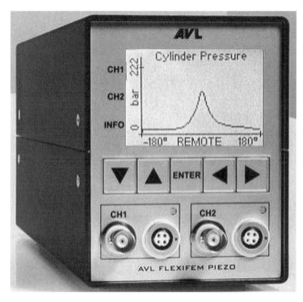

zero-level correction is to measure the inlet pressure with an absolute measuring transducer. An example of this evaluation-and-trigger logic is shown in Figure 5.21.

Additional Evaluation Possibilities for Intelligent Amplifiers

Combustion Noise   Evaluation of combustion noise from diesel engines is a commonly applied technique in the calibration process of the engine electronic controls system. The noise generated by the combustion event itself is an optimisation parameter and target for producing a smooth, refined diesel engine. The details of this evaluation are discussed in a later section of this book. However, this measurement is ideally suited to being integrated into an intelligent amplifier concept. For many years, AVL has produced a stand-alone combustion noise meter that takes the form factor of an amplifier but it does not include an amplifier function in the device itself. The latest generation device that includes amplifier function (an intelligent amplifier) is shown in Figure 5.22.

The combustion noise algorithm is simply an advanced filtering process on the raw measured curve. It is established in the time domain, which means that no crank degree reference or encoder is necessary. This is ideal for integration as an intelligent amplifier function and can easily be combined with the charge amplifier and evaluation unit in a

Figure 5.20 Peak meter for large-engine condition monitoring
(Source: Lemag.)

single package, analogous to the peak meter discussed above. Calculation of the noise value on a cyclic basis and output of this value as a voltage for measurement on external equipment will facilitate a low-cost, flexible package for this specific application.

Knock Meter   Measurement and establishment of the threshold of engine detonation as a function of ignition timing is a standard evaluation for calibration of gasoline engines. The fundamental evaluation involves measuring the cylinder pressure curve, applying a high-pass filter, then extracting the maximum, peak overpressure value due to the knock event on a cyclic basis. This value is then often used in further statistical evaluations to determine the frequency and intensity of knocking, and for establishment of an absolute threshold of knock/no-knock condition at the knock borderline. This evaluation technique is also suitable for integration into an intelligent amplifier, in that the extraction of this result from the measured curve consists of straightforward filtering and calculation. A more sophisticated statistical analysis and processing of this knock overpressure result could be done internally inside the unit with a more advanced processer and appropriate parameterisation. The results of this evaluation could then be output through analogue or digital signals. The main target, though, for a knock meter is to protect the engine in the absence of a full combustion measurement system, to save costs and installation effort (monitoring rather than measuring!). The threshold of knock overpressure would be user definable in the user interface of such a device (via parameters) such that when a

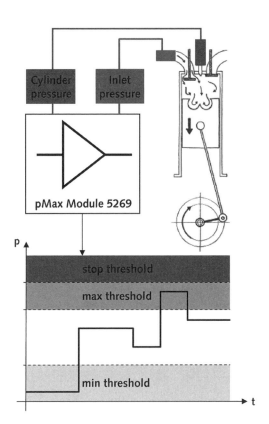

Figure 5.21 Evaluation of peak signal and trigger thresholds
(Source: Kistler.)

knock event occurs, a discrete signal from the device could be connected to the test bed controller to react immediately and prevent engine damage. One disadvantage, though, is that, for this knock evaluation, an encoder signal is needed to establish crank position for the measurement window. This does not need to be highly accurate, however, so a simple encoder or sensor could be employed.

IMEP Meter For combustion pressure measurements, the most fundamental task it has to be able to establish the cylinder work output. This is required for many test applications, particularly where engine friction effects are to be established. In addition, many in-vehicle calibration tasks require an engine torque–related signal, and indicated mean effective pressure (IMEP) is an ideal basis for correlation with torque/IMEP data from the engine test bed. A common requirement is that the IMEP should be calculated, organised, and displayed in firing order. The result of this organisation gives an IMEP trend curve, over all engine cycles and cylinders, that has a direct correlation with an instantaneous engine torque signal. This is a particularly useful derived value for in-vehicle calibrators, where engine torque cannot be measured directly during vehicle tests (due to the powertrain installation in the vehicle). This functionality could potentially be executed in an intelligent amplifier concept.

A possible system model would use each single channel amplifier module to calculate the IMEP result for that cylinder in real time. An additional processor would then collect all

135

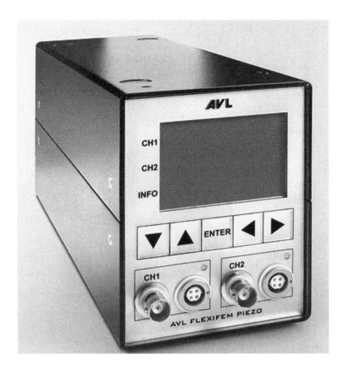

Figure 5.22 Combustion noise meter
(Source: AVL.)

the result outputs from each module, then synchronise and organise them for display and output. The most important function would be the possibility to communicate the IMEP "parade" data set as an output in real time for recording externally, which could be done via a voltage digital to analogue converter (DAC) output , or via a controller bus network (CAN) bus interface on the system. In this way the signal can be recorded on external time-based recorders for correlation with other measurement channels or devices.

The main challenge for correct IMEP calculation is to ensure correct top dead center (TDC) position establishment. This requires an accurate encoder signal, so an IMEP meter would need an accurate method of recording crank degrees and establishing TDC correctly. These are key features of a combustion-measuring system, so there is a question as to whether an IMEP meter should be a sophisticated amplifier or a simple combustion-measuring system. This issue would have to be resolved in the early development phase of such a device.

5.1.6 Summary—Signal Conditioning

There are many measurement technologies for scientific and engineering environments that migrate into engine development and the test environment. Most manufacturers of combustion-measuring equipment provide charge and analogue signal amplifiers. Less common technologies have been used and developed into commercial amplifiers for combustion measurement, but there are always compromises in migrating laboratory equipment into the engine test cell, which is a harsh environment for sensitive electronic equipment. When examining alternative measurement technologies for combustion measurement procedures, it is worthwhile to examine the market for available signal-conditioning technologies; these should be given preference. Novel measurement techniques, unsupported by commercial manufacturers, are not productive and should be confined to the very early stages of the engine development process in a fundamental research environment.

5.2 Measurement Hardware—The Data Acquisition and Measurement System

5.2.1 Introduction and Overview

The measurement system hardware is the heart of a modern, digital combustion measurement system and provides the "intelligent" part of the measuring chain. It records, in electronic form, the inputs from the signal-conditioning chain (i.e., transducer and angle encoder). This information is stored digitally as vector data sets with the correct abscissa and hence domain (angular). The stored data are subsequently processed to provide, as quickly as possible, the calculated results derived from the raw data that the engine engineer or control system needs. This has to be transferred to the user, normally via a personal computer interface, for visualisation and storage. In addition, the measuring system hardware must communicate with other connected systems to allow intelligent and efficient transport of combustion-related data to external control and data acquisition devices.

Most commonly, the system core is a processor board or boards, interfaced via fast digital communication networks both internally and externally. The communication and internal handling of the data sets is critical to overall system performance and capability. Fast, modern processors are easily capable of this task, but optimisation of the real-time process handling is an essential element that is dictated by the overall software design and development model, which must produce optimised, deterministic operating software to support the measuring task requirement.

5.2.2 Operating Requirements

The basic requirement of the measuring hardware is to convert the system inputs into usable combustion-related data or parameter sets for the operator or connected subsystem. Generally the system operates in one of two modes:

- **Monitoring (oscilloscope) mode**: continuously running and recording data in a buffer, with screen updates online to display data.
- **Measuring mode**: recording of a predefined number of engine operating cycles. Once this is complete, the operation is stopped and those cycles are available for analysis or permanent storage in a data file format.

The simple measuring task can be broken down into a measurement data-flow sequence:

1. System inputs are predominantly angle encoder and cylinder pressure (or other crank angle–related subsystem such as injection activity). The angle encoder crank degree and trigger marks are fed into the system as dedicated input channels. The cylinder pressure inputs are analogue voltages, and these are normally fed into the high-speed analogue inputs, generally at least one per instrumented cylinder.

2. The inputs are processed by the system. From the angle encoder, the cylinder volume is calculated in real time from the angle marks, and this requires engine parameter information to be input into the system via the user interface. The analogue inputs are digitised via dedicated or multiplexed analogue-to-digital converters at an appropriate resolution, and triggering of the sampling is effected via encoder pulses. Scaling factors for the analogue inputs are also stored into the system as parameters, by the user, prior to the measurement. These fundamental inputs are measured in real time; each raw sample can then be referenced to a physical value (e.g., pressure in bar, and crank angle position).

3. The data samples are written, in real time, into the system dynamic memory. This allows immediate storage of the acquired data set and is an essential feature of modern combustion measurement system. Clearly, this process is limited by the size of the system memory, and this depends on the number of data points per engine cycle (i.e., the crank degree measurement resolution). Another limiting factor for the hardware is the speed at which each sample can be digitised, also known as the throughput rate. This depends upon the speed of the analogue-to-digital converter, as well as the engine speed and the crank degree measurement resolution. The general relationships are

$$\text{System Throughput (kHz)} = \frac{(\text{Engine Speed} \times \text{Channels} \times 0.006)}{\text{Measurement Resolution}}$$

$$\text{Required memory (MBytes)} = \frac{\left(\frac{\text{samples}}{\text{cycle}} \times \text{Engine cycles} \times \text{Channels}\right)}{500{,}000}$$

Units: engine speed (rpm), measurement resolution (crank degrees). Note: if analogue-to-digital converter is not multiplexed, then channels = 1.

4. In many systems, during acquisition of the data in memory, a system signal processor or dedicated system processing resource has access to the raw data sets for the purpose of calculating result values. This optimised process allows the creation of additional data set vectors containing cyclic-based data sets of calculated values of interest from the raw data—for example, the maximum value of the pressure curve or the IMEP for that cycle. This is information needed to understand the combustion process and perhaps react to it (in a control loop). The data sets are stored in memory such that each value can be correlated with a cylinder and cycle number.

5. The data stored in memory is transferred to the operator user interface, which in most cases is a personal computer. Generally, the most recently acquired cycle is transferred and displayed on the screen. This is continuously updated (overwritten by the latest cycle) as long as the measurement is active, giving a display object similar to an oscilloscope. In addition, results are also transferred and displayed with the option of generating statistical values from the result data sets.

6. Once the measurement is completed, or the monitoring oscilloscope display stopped, the full crank angle data sets for each cycle are transferred and merged with the results into a file-based format that can be written to a hard disk. Note that if the system has a personal computer as the user interface with a connection to the measurement system

via a proprietary link (Ethernet, Firewire, USB), then this process can take some time, depending on the size of the data file (a function of measurement resolution, number of cycles, number of channels, etc.). This factor may impede the measurement task because, in general, it prevents the restarting of measuring or monitoring until the data are completely transferred out of the system memory.

7. Once the data are transferred into a file, they can be saved for future analysis or reference. The system is then restarted to monitor or measure again.

5.2.3 System Interfaces

The general inputs and outputs found on a modern digital processor–based combustion measurement system can be summarised as follows:

- Angle (shaft) encoder
- Operator interface
- Analogue signal interface
- Digital signal interface

5.2.3.1 Angle Encoder

Normally a dedicated input socket or connectors for trigger and crank degree marks from an incremental encoder. Generally TTL inputs but other technologies (LVDS, RS485) can also be encountered. It is also commonplace that some level of angle encoder signal conditioning is built into the system. This is often in the form of a simple pulse multiplier to increase the resolution of the incoming encoder crank degree pulses up to the required level; this is generally no greater than 0.1 degree crank angle.

5.2.3.2 Operator Interface

Most modern systems employ standard personal computer technology as a platform for the user interface because this is familiar, generally utilised technology. The speed of this interface is a decisive factor in certain aspects of the system performance, for example, screen refresh rate and data-saving time. Typical interfaces employed are Ethernet and Firewire; in addition, fast serial interfaces like USB can be employed.

5.2.3.3 Analogue Inputs and Outputs

The analogue input channels are the high-speed channels used for measuring the combustion-related parameters related to angular position of the engine. Generally, they are the interface points for the signal-conditioning systems. The analogue voltage signal from the signal-conditioning amplifiers is connected at this point and transfers the measured channel of interest into the system from processing. Common interfaces are made through standard connectors, for example BNC or D-SUB. The analogue input consists fundamentally of a high-speed digitiser or analogue-to-digital converter. It samples the input at regular crank degree intervals (according to the measurement resolution) and converts each value into a digital bit for transfer and storage in the dynamic memory of the system.

The Measurement Chain: Measurement Hardware

ADC Bits	Min. pressure change (bar) at 50 bar range	Min. pressure change (bar) at 100 bar range	Min. pressure change (bar) at 200 bar range
12	0.01221	0.02441	0.06104
14	0.00305	0.00610	0.01221

Figure 5.23 A table showing minimum pressure change measurable according to ADC range.

The resolution of the analogue to digital converter is an important factor that has a decisive affect of the quality of digital conversion and subsequent calculated and stored data. It determines the minimum amount of pressure change that can be recorded. This value can be determined by:

$$\text{Minimum Pressure change} = \frac{\Delta P}{2^x}$$

where ΔP = Pressure range applied across ADC range and x = bit resolution.

Note that the resolution and quality of digital conversion have an inverse relationship; therefore, the analogue-to-digital converter chosen by the system manufacturer will be a compromise based on these factors as well as cost. It is common to find 12-bit or 14-bit converters on current systems because these can provide high-quality conversion with an appropriate quality and speed. The comparison between 12-bit and 14-bit conversion with respect to pressure range and minimum pressure change detectable is shown in the table in Figure 5.23.

Note that certain systems may share a converter across channels; that is, the input to the converter is multiplexed. This has an effect on the total system throughput rate, which is reduced dramatically as the channel utilisation on the system is increased. However, most modern systems incorporate a separate converter for each input channel in order to avoid this restriction.

Figure 5.24 A knock cylinder pressure signal highlighting dynamic range requirements.

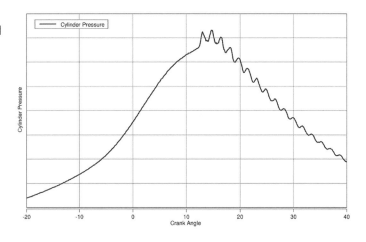

The bit conversion resolution affects the quality of the measured data, and it is particularly important to consider this effect where the signal of interest has a large dynamic range. An example would be measurement of a cylinder pressure signal that contains high-frequency knock components that are of interest, as shown in Figure 5.24.

In this case the converter input range has to be scaled between the maximum and minimum of the complete signal, although the high-frequency part has a peak-to-peak range that is only a proportion of the measured total signal range. Therefore, the number of digital bits for this part of the signal must be sufficient to capture the high frequencies with sufficient quality, even though the total number of digital bits has to be "shared" across the total signal range. This factor supports the requirement for high–bit resolution converters, and if the system in use does not support this, it may be necessary to pre-filter the signal, reducing higher amplitude, lower frequency components, and thus zooming the converter range into the area of interest on the measured signal. For general-purpose signal acquisition this is an acceptable workaround, but for combustion measurements the whole signal range is of interest, so relatively high dynamic range is required.

Generally, 14-bit is the maximum dynamic range found on commercial combustion measurement systems as this provides an appropriate compromise of performance and price. Commercially available, multipurpose processor–based acquisition boards can employ converters with a higher dynamic range, and these can be programmed to acquire combustion data by triggering acquisition via an external clock pulse (from angle encoder marks). These can be in the range of 16 to 24 bits, but conversion quality, dynamic range, cost, and speed are all compromising factors and the typical specification chosen by manufacturers of combustion measurement systems should be considered when choosing a general acquisition board for combustion measurement tasks.

Certain applications—for example, noise-vibration-harshness (NVH) or acoustic measurements applications—require very high digital conversion quality because the signals have high dynamic range requirements (at least 24-bit). Note that for combustion measurement applications, this is not generally required.

In general, the voltage range of the analogue inputs is fixed (–10 to +10 V) although some systems and card manufacturers offer programmable input gain. It is usually the task of the signal-conditioning system and setup of that system to ensure that the analogue input range is optimised—that is, as many as possible of the available bits are utilised for signal conversion (minimal distance between digitisation steps)—and as a general rule, the input range voltage should be at least 75% of the converter range voltage, allowing a small margin to ensure that any slight drift does not cause the signal to drift out of the upper or lower limit of the converter input range.

Outputs are commonly analogue. This allows the output of a result value to be expressed as an analogue voltage at an output connector and can therefore be a useful way of transferring data from the combustion measurement system to an external recording or monitoring device. Combustion measurement systems typically offer between 8 and 16 analogue output channels, in the range of –10 to +10 volts, the input scaling factor being set appropriately to the range of the result output. For example, a maximum pressure result may be in the range of 0–200 bar. With an output range of –10 to +10 volts, this gives a transfer function of 10 bar per volt, so a value of 120 bar would appear at the analogue output as +2 volts. The analogue output-processing card is mounted directly in the hardware unit and generally has access to the real-time results, directly from memory. This means that the response of these outputs is generally deterministic, and the results can thus be transferred to a data acquisition system or control system that requires real-time result information from the combustion measurement system for engine optimisation or development tasks.

5.2.3.4 Digital Inputs and Outputs

Because combustion-measuring systems are high-speed acquisition devices, it is common that they are also equipped with digital inputs and outputs. In addition, these are commonly found as part of the input/output of general-purpose acquisition cards incorporating analogue and digital channel handling and processing capability. The simplest application is to measure individual bit status against crank angle; this is an efficient way of measuring a signal that has only one of two states because the data set size per cycle is greatly reduced when compared to the analogue input. One such value is that of an ignition coil driver signal: from this simple on/off signal, coil charge and ignition timing can be derived. Often it can be useful to measure a complete bit pattern against crank angle and display the numerical value. For example, 8 input bits can describe any number between 0 and 256. This could be used to display the bit pattern as a number from a parallel port–type device output.

Another common application for digital signals is counting time difference between edges with a timer-counter applied to an incoming digital signal. A common application is to use this function with angle encoder signals to determine instantaneous speed of the engine for further calculation of engine rotational vibration and dynamic torsion between two points on a rotating shaft.

Simple digital outputs can be used as triggers from a specific event. For example, digital signals can be employed to change state when a threshold is reached or exceeded and can therefore be used to signal a fault condition to an external monitoring or measuring system for reaction.

Generally, the physical interface connection is by standard industrial connector technology, usually mounted on the front or back side of the measurement system. In all cases, the manufacturer's information should be used to determine the specific requirements.

5.2.4 A Typical System

Figure 5.25 shows the layout of the hardware of a typical, modern combustion analyser system. It is possible for the signal-conditioning hardware to be built into the hardware rack; this is useful for an instrument that must be portable (e.g., a unit shared between cells) or used for in-vehicle measurements. The optimum system layout at a fixed, test-bed installation would be such that the amplifiers are remote and mounted near the engine, as close to the transducers as possible. This reduces the possibility of noise induction on the low-level signal lines. Analogue cables transmit the amplified, conditioned signals from the amplifiers to the measurement hardware.

The angle encoder is mounted directly at the engine, and optical-to-electrical signal conversion should ideally take place away from the encoder head. In addition, the digital

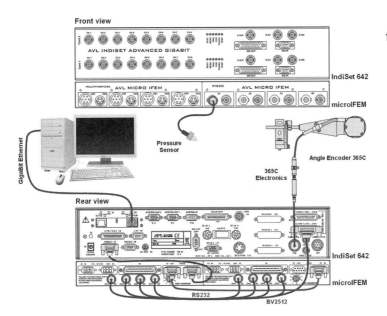

Figure 5.25 A typical combustion system hardware measurement environment
(Source: AVL.)

signals from the encoder should be transmitted as differential signals to increase common mode rejection and reduce the possibility of noise on the signals.

The measurement system is normally housed in a rack that can be mounted in an industrial-type frame or cabinet. Alternatively, it can be mounted on the bench top. Typically the hardware itself runs from main voltage and includes a wide-range power converter to reduce the effect of supply voltage fluctuation. In addition, many systems offer the option of low-voltage-power converters so that they can operate at low DC voltage levels for in vehicle use. Often the cabinets include an instrument ground connection in addition to the protective ground connection. If this is available, it should be used to reduce the effect of any ground loops or differences in ground potential.

The interface from the system to a computer-based user interface is generally accomplished with a commercial personal computer–based interface technology: Ethernet and Firewire are very commonly used, as these are high-speed interfaces with well-understood protocols for software developers to utilise and optimise. If the combustion system is a computer-mounted, processor-card type, then USB will also be encountered. It should be noted that acquisition card-based systems are generally considered as laboratory-based equipment and, as such, may not survive the test cell environment, which is generally considered to be an industrial environment and is dirty and noisy, in both physical and electrical contexts. The preferred solution is that the processor board be integrated in an industrial rack mounting system for durability and noise immunity. It should be noted that if a personal computer is used as the man-machine interface, then the same general guidelines apply as for the measurement system; that is, an industrial rack mount machine should be used where possible.

Data storage can be via the system computer hard drive, but in large test cell environments, it is common that data from the measurement system have to be transferred to a central host. This can be easily integrated using standard personal computer network interfaces, tools, and protocols. Generally, in addition to data, parameter files that dictate the equipment set-up configuration are also backed-up centrally.

Chapter 6

The Measurement Chain: Measurement System Software

6.1 Software—The User Interface

6.1.1 Introduction

Current digital combustion measurement systems most commonly use a personal computer as a user interface. Today this PC based technology is generally familiar to most technicians, engineers and scientists who are likely to use or need a combustion measurement from an engine. This requires that the user interface software be developed in an environment so that the features and functions of the combustion measurement operating software have a look and feel similar to the computer operating system (normally Microsoft Windows™).

It is worthwhile to note, though, that older systems, developed before the popularity of personal computers, have a specific operating system and user interface that may not be familiar to a PC user. Even though these systems look unfamiliar, the user interface and operating system were optimised for the task, and an experienced operator could execute measuring tasks with great speed and efficiency.

The advantage of using a PC based operating system is that development environments and programming technology for PC based software are readily available, in conjunction with software engineers who know how to create user interfaces that are easy to understand and workflow oriented. In addition, network technology and high-speed interfaces are available on a modern PC, and they can be utilised for easy interfacing with the measurement hardware and centralised data servers for fast transfer and storage of measurement data files.

The basic function and requirements of the operating software and user interface are presented here:

1. Parameterisation and configuration of the measurement system before a measurement. This includes setup and definition of the measuring task and operating mode, setup of input and output channels and interfaces, and entering of engine parameters and other test-specific information

2. Operation of the system during the test, either by manual user intervention, or via remote control from an external control system. Typical tasks would be to start monitoring, start measuring, save data, change system setup parameters, transfer results and parameters between combustion measuring system and external control and data acquisition system, and to manage communication interfaces, including error handling.

3. Allow the user to create display objects and windows so that data can be easily and clearly displayed during run time, while a measuring or monitoring procedure is executed. In addition, it must be possible for premeasured data to be easily viewed and displayed after a measurement, for postprocessing and analysis.

4. Allow the user to save data during a measurement, then easily access and reload it after a measurement procedure. In addition, allow the reloading of historic data for further analysis or comparison. Facilitate the export of data (including parameters if required) to external programs in various appropriate formats for further processing, analysis, or display.

Additional features may be required according to measurement requirements and the functions and options included on the system to be used.

6.1.2. User Interface

Generally, the main operative tasks when using the combustion measurement system fall into 3 main tasks:

1. System setup
2. Measurement execution
3. Data evaluation

The software normally consists of elements or subprograms to deal with each of these main tasks. An explanation of the main operational areas follows here.

6.1.2.1 Parameterisation

A number of basic parameters must be set correctly prior to a measurement. These relate to equipment setup, the measurement task to be undertaken, and information about the unit under test (i.e., the engine, in most cases). Generally, this information is stored in a parameterisation file that can be saved after a setup configuration has been completed. This arrangement allows reloading of setup information when a task has to be repeated (saving time). In addition, existing parameter sets can be adapted rather than remade from scratch for new applications. For example, it may be necessary to change engine parameters only due to a change of unit under test, as opposed to starting with a new, blank parameter file that would need to be populated with data. This procedure improves the efficiency of the parameterisation task.

The main parameters to be addressed are listed here:

Engine settings—Bore, stroke, conrod (connecting rod) length, compression ratio, firing order, etc. These data are particularly important because they are used in the calculation of results, which the system generally executes in real time. Additional test system or engine parameters may also need defining for saving in the measurement data file: for example, values transferred or recorded on other data acquisition systems, e.g., manifold pressure, throttle position, engine temperature, and fuel type. These values may be important for reference or for further extended calculations involving the combustion-related data.

Angle encoder settings—Definition of the number of incoming marks per revolution, plus setup of input signal types (analogue or digital). Definition of output marks (hence multiplier settings). In addition, test dialogue to count marks, and display values to assist the user in setup and faultfinding when encoder-related problems occur.

TDC settings—The correct definition of TDC is essential for the system to calculate certain result values accurately. The user interface should support and guide the user through the process of establishing the absolute position of engine TDC relative to the position of the angle encoder trigger mark so this information can be stored in the parameter file. Note that TDC position should be rechecked periodically by the user during testing. Hence the user interface must facilitate this process so that it is easily executed on a routine basis. In addition, the software must support various methods of TDC establishment according to commonly used and established methodology (this topic is discussed later in the appropriate section).

The Measurement Chain: Measurement System Software

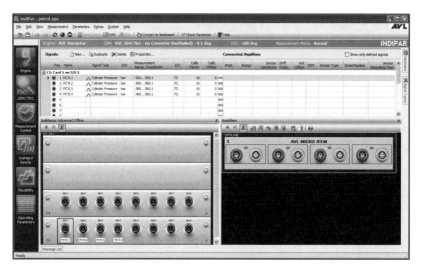

Figure 6.1 A typical parameterisation display dialogue in the user interface software.
(Source: AVL Indicom.)

Channel settings—Each measurement channel used needs to be appropriately configured. This involves naming the channel, positioning the channel on the measuring system, making calibration settings (gain and offset, zero-level correction method), measuring range and resolution, and calculating or deriving results from the raw data curves.

Measuring sequence—Definition of the measuring task, number of cycles, averaging, activation of calculation execution, data-saving path and sequence, and execution of additional processing scripts or sequences. A typical parameterisation display is shown in Figure 6.1.

6.1.2.2 Display of Data

In order to view the recorded and calculated data during a measuring sequence, display objects must be arranged according to the user requirements or preference. This must be a completely flexible part of the software user interface. Typical systems generally provide a number of standard graphical object types that users can select and integrate into their specific display configuration. Most systems support multiple page displays. Each page can display a number of objects so the user can define the relevant graphical objects to display, according to the measurement target or task. A common feature is that the user can lock the display (via user level login to the software) so that, once configured, the display cannot be accidentally spoiled by inadvertent use of the operator controls (for example, the PC's mouse).

Each display object generally has its own configuration dialog in which colours, styles, scale range, etc., can be adjusted. In addition, the user can select or exchange data channels for display in that specific object. In common with Windows applications, it is usual that display objects can be copied, cut, or pasted between display windows, and this capability accelerates the development of standard display windows that can be saved within the parameter file. In addition, display definitions can generally be saved separately and imported or exported to and from parameter files to give the user a high degree of freedom and speed of development of display configurations.

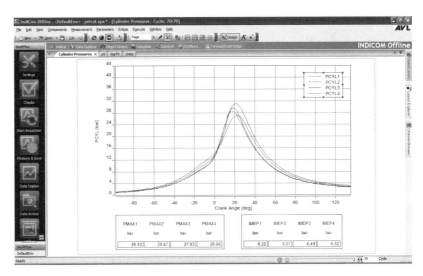

Figure 6.2 Online data display.
(Source: AVL Indicom.)

For general operation and manipulation of the display, particularly for analysis during measurement, additional display features are incorporated to assist the user. These typically comprise cursors (differential and band type) as well as zoom and shift features, so that measured and calculated curves can be examined in detail. A typical online data display is shown in Figure 6.2.

6.1.2.4 Data Management

Data management is an important factor that dramatically affects workflow efficiency. Generally, the user interface allows access to historic data so it can be easily reloaded into an existing display configuration for analysis. A useful feature is to be able to display historic data in the same window as current measured data, allowing a file comparison of measured data against reference data.

Measured data files are generally in a specific binary format, often including parameter information embedded within the file, facilitating result calculations. The data required to generate the volume channel, in addition to the measured pressure channels, are present in the file structure, in addition to calibration factors, operating parameters, etc. The calculations can thus also be done in external programs that can read the source file format.

Generally, comments can be stored in the measurement data file. These could be manually entered by the operator (after a measurement) or automatically entered (via an external digital interface) to an automation system. It is useful to note that there is no standard file format used by combustion data system manufacturers. Each has its own native binary format. In addition, there is no standard data model, as is the case for other engine test bed measurement and automation systems. These standards are proposed by standardisation bodies. In Europe, for example, ASAM (the Association for Standardisation of Automation and Measuring systems—www.asam.net) provides this function.

Many systems also offer text configuration files that document the hardware setup. Such files can be useful for retrospectively understanding how the system was configured for a particular measurement task that generated some data that is of interest or that is under analysis.

Saving data normally requires correct setup with respect to the file path. The software interface generally provides this function, and, in addition, standard naming conventions and automatic incrementing of file extensions are common features to ensure that files can be searched according to logical criteria, and that files are not overwritten during a measurement series.

Once data are saved in the required directory and can be accessed from the user interface, postprocessing of premeasured data can take place. If this process is executed at the measurement system, the device is not available for measurement tasks, and that arrangement reduces efficiency. Often, simple data postprocessing software is supplied with the system for use offline (i.e., as stand-alone software). This software can be installed on a separate PC in an office environment, allowing the engineer to process data while the operator executes measurements. This arrangement can provide a particularly efficient working environment if the data are saved to a central server location that both have access to.

Often this software is supplied with the system, and may be distributable to other users in the test environment. The interface is similar to the operating system (generally an offline version), and it allows visualisation of premeasured channels and precalculated results. Often these tools can be used as a source of display objects for presentations or documents, as the displays can be exported as image files (jpg, bmp, etc.) to be imported into documents. Note though that generally, further calculations or extended calculations on the measured data are often not possible with these tools. Generally, for this requirement, full postprocessing and analysis tools are required. These can read the source data and parameters and provide the possibility to define processing algorithms via script or formula programming languages available within the software, or via commonly used external scripting programs, for example Microsoft Visual Basic™ or Matlab™. These advanced software tools are generally purchased additionally, and they are not part of the combustion system software scope of supply.

A requirement within the user interface software is the ability to export data, as mentioned. Exporting text-based files that state the system configuration and exporting OLE (object linking and embedding) display objects for use in external desktop programs is common. Often, though, it is required to be able to export the data files, or to save the raw data in a nonnative format. Common formats required are delimited, column-oriented text files, as these can be readily imported into well-established desktop analysis tools (for example, Matlab™ and Excel™). It should be noted, though, that importing and handling cyclic crank angle based multiple-vector data sets in these programs can be difficult and time-consuming. In such cases, data averaging and the possibility to create data set envelopes from the measured curves are useful features to be able to represent the measurement with a reduced number of data bytes. Another useful feature is the possibility of reducing cyclic result values to statistical data tables. It is a useful data reduction technique that does not greatly affect data quality for the end user and that simplifies the handling of combustion data in generic, external programs that are not specifically designed for the task.

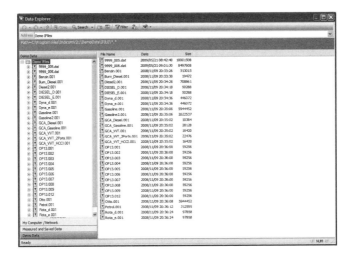

Figure 6.3 Data access tool within user interface of combustion measurement system software.
(Source: AVL Indicom.)

Often data processing, reduction, and conversion tasks can be executed as batch jobs if the combustion system software has the capability to execute user-defined scripts. If this is not the case, the external desktop programs mentioned above very often have this capability. This allows the user to select a batch of files for reduction and/or analysis, and then allow the software to execute this task over a number of files, facilitating an efficient data analysis process in the work environment. In addition, it would be possible to include plausibility-checking criteria in the automated process to identify and highlight rouge data points in measured data files. Figure 6.3 displays a data access tool user interface for combustion measurement system software.

6.2 Features and Operating Modes

6.2.1 Standard Measurement Operations

Fundamentally the combustion measuring system has two main operational modes (as mentioned previously):

1. *Monitoring or processing mode*—Also known as the oscilloscope mode, the system continuously measures and displays data on the user screen, continuously updating the display with the current cycle. The system generally will have a cycle buffer so that when this mode is stopped, the user has access to a history of cycles (typically a hundred or so, depending on equipment memory size).

2. *Measurement mode*—The system records a predefined number of cycles, and then stops measuring. Data are held in a dynamic memory buffer for immediate analysis by the user; the data can be saved (via transfer to hard disk as a file), either manually or automatically. Often it is also possible to implement data reduction (manually or automatically) prior to data saving to reduce file size.

These two modes fulfil the requirements of most standard cylinder pressure measuring tasks. In addition, other operating modes are available for system calibration and parameterisation:

Figure 6.4 Deadweight analysis parameters.
(Source: FEV CAS.)

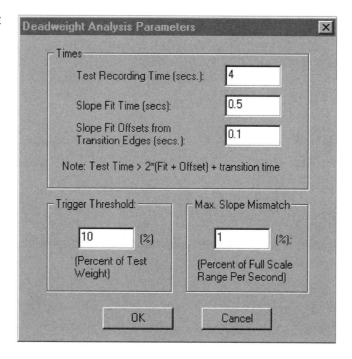

1. *TDC find*—The system measures a curve generated by an engine-installed sensor then applies an algorithm to find the correct relationship between the angle encoder trigger mark and the actual, physical engine TDC position under dynamic conditions. Once established, this information can be stored in the parameter file (discussed in detail in the appropriate section of this book).

2. *Calibration mode*—Often, with the piezoelectric measuring chain, a complete calibration of all components in the chain is required for accuracy and repeatability. Due to the nature of operation of a piezoelectric sensor (dynamic rather than static), a pressure jump test using a dead weight tester is often used, and this task has to be implemented in the software via a user-guided interface. This helps the calibrator execute the pressure jump in the correct way. It also helps provide some data averaging before and after the pressure jump, serving to stabilise the measuring procedure. Often included in calibration mode is a dialogue showing input values in bit steps and volts for each channel, to assist in channel calibration of analogue inputs. As systems have evolved, however, and based on user requirements, additional modes of operation are available for specific measuring tasks. Figure 6.4 portrays a deadweight analysis parameters window for Windows-based calibration software.

6.2.2 Special Measurement Modes

Additional measurement applications have been conceived over the years as digital combustion measurement systems have evolved in sophistication and technical capability. In addition, as engine technology and engine control systems have become more sophisticated to cope with performance demands and emission constraints, more emphasis has been put on the use of a combustion measurement system to understand normal combustion events and identify abnormal combustion phenomena. This has promoted the development in the software user interface (and hardware) of special measurement tasks or features. Typically, additional tasks included in the software configuration of modern systems would be these:

Knock measurement—This is a very common operating mode or feature available on most (if not all) combustion analysis systems, either as an option, or as part of the standard scope of delivery for the system. It applies to the measurement and analysis of abnormal combustion conditions in spark ignition or gasoline engines. Knocking combustion is caused by the premature, uncontrolled, self-ignition of the fuel-air mixture shortly before it is reached by the flame front of normal combustion. The cause for this instantaneous self-ignition of the end gas is often the rise in pressure in the entire combustion chamber during combustion, which leads to a temperature rise in the end gas. In extreme operating conditions, this temperature rise provides sufficient thermal energy to spontaneously ignite the end gas, before the controlled flame front, advancing from the spark ignition event, reaches it. This undesirable phenomenon can be encouraged by poor combustion chamber design, combustion chamber deposits, or insufficient localised combustion chamber cooling.

There are various algorithms in use, and these will be discussed in the appropriate section of this book, where knock measurement will also be explored in more detail. An important point, though, is that this operating mode requires an intensive, real-time processing capability of the system because the raw cylinder pressure curve must be filtered and evaluated according to the applied algorithm, such that abnormal combustion events can be identified and reacted upon immediately at the end of that same engine cycle. Often it is required to output the knock event yes/no signal, via an appropriate interface, such that it can be measured on the engine or test system controller, so that appropriate action can be taken to prevent further or excessive damaging knock. This is a typical requirement in an engine-calibration environment where protection of a prototype or base engine is very important, particularly if the engine-control system has no, or only partially, populated calibration maps.

Cold start—This function is often required for detailed analysis and understanding of emission-generating combustion events when the engine starts from cold. The measurement system must be able to start acquisition right from the very first movement of the crankshaft. Of course this is not really a technical challenge; the problem is that, although the data can be acquired, it can not be properly referenced to the correct TDC position, as the engine phasing and position is not known until a full, measured cycle is complete. Therefore, the system must start to acquire from the first crank mark it receives, then, after the measurement has stopped (after an appropriate number of engine cycles), the data must be shifted correctly to give the correct relationship of the measured curves relative to the correct TDC position of the measurement table. Once this has been done, all appropriate calculations must be executed for each cycle to return the appropriate results for analysis. Note that it is very often the case that this cold start data must be available in the time domain as well as in the angle domain. This is to allow data correlation with other measuring systems (typically fast emission measurement devices) that normally measure in the time domain. Clearly, the angle domain is still required for correct calculation of results, and hence the system must support both time and angle domains in this mode of operation.

Event mode—This is a requirement where the system will be used to capture data on the occurrence of a specific event. Typically, this would be an engine-monitoring activity that may be implemented to protect an engine during an automatic test run on an engine test bed. In this mode, the system can be set to monitor specific result values to ensure that they lie within a user-defined tolerance level. An example would be to monitor a diesel engine to ensure that the cylinder peak pressure did not exceed a mechanical design limit value.

The system can be set to monitor the value from a single cylinder or from multiple cylinders. Then, if the limit is exceeded, a defined action can be initiated. An example example of this would be a digital signal state change that is connected to the engine or test bed controller so that an appropriate, predefined emergency reaction can be initiated. This monitoring can be combined with a data-saving process to store pre- and posthistory around the event for postmortem analysis and to facilitate an understanding of any fault condition by the engineer. In addition, many systems support combinational logic that can be applied when one is monitoring single or multiple results. This capability improves the plausibility of the reaction. For example, an event must occur a certain number of times, within a certain number of cycles, before that event has been deemed to have occurred, and a signal is then output.

Certain available systems have sufficient capability to continuously monitor events while executing other tasks. This means that with such a system, the engine is always protected and monitored, even when the system is measuring and saving data (at the same time). This is a particularly useful feature in an engine calibration environment where a DOE (design of experiment) test is being executed. In these conditions, it is often the case that engine steady state measurements will be executed, requiring a data save process that takes finite time, while the engine will be operated transiently in design space to the next measurement condition (speed and load point). If this is done when the engine controller is not fully calibrated, and hence the engine's response to speed and load in this condition is unknown, there is a great risk of abnormal, uncontrolled combustion phenomena that can run away and damage an expensive, prototype unit under test.

Time-based mode—Combustion analysis equipment is generally a high-speed digitising device, triggered by encoder pulses. All systems offer the capability (as standard or as an option) to operate in time-based sampling mode as a true, time based measuring system. In this mode the analogue-to-digital sampling is triggered via an internal clock/oscillator instead of by angle encoder marks, and the data are recorded in the time domain, producing dynamic curves of each channel, in a style similar to that of a chart or strip recorder. This mode is occasionally required for combustion measurements where analysis of trends and comparison of combustion data with time-based data from other recording systems is required. In addition, it may be necessary to record combustion pressure data in the time domain in order to correctly execute certain further calculations for data analysis. An example is where an FFT (fast Fourier transform) is to be executed when time-based data are required in order to be able to correctly establish the frequency components. Note that in this mode, only raw data recording is possible; generally result calculations are not enabled in this mode.

A more useful time-based measurement mode is one in which the system supports simultaneous angle and time domain recording, with the possibility of correct correlation of the data in the raw data file format. This is a common feature on more sophisticated, high-end systems that have multiple data acquisition cards. Often simultaneous sampling of time and crank angle data (synchronised) is used in applications such as cold start measuring. It should be noted that high-speed, time-based channel recording creates large data sets that have to be transferred to the user interface (normally a PC) in real time (as opposed to the crank angle data, which are cyclic. For crank angle data, normally, only the current cycle is transferred and redrawn on the screen. Also, crank angle measured data normally have a much lower throughput rate than the AD converter is capable of). This can create performance limitations due to the available bandwidth of the connection between the PC and the combustion measurement hardware. This should be considered when using this mode, and if high-speed recorder sampling is used, display of the data during measurement run time may not be a practical possibility.

Torsion and rotation analysis—This operating mode uses the digital timer counter function of the digital and encoder inputs on the acquisition card with respect to the incoming marks from the engine angle encoder, or from other encoders fitted to the unit under test. The system counts the time difference between the incoming encoder teeth. The distance between these teeth is preparameterised in the setup of the system so that if the distance between the teeth is known (in engine degrees) and the time difference is measured (by the timer counter), and then the instantaneous speed variation within an engine cycle can be measured. If this speed variation is compared with the average speed per cyclic degree on a crank angle basis (for example, a measurement sample at each engine degree), then the difference is the deviation between the theoretical position (based on uniform rotation within an engine cycle) and the actual position. These data can then be resolved into the angular domain, giving a deviation curve (in degrees) of the actual engine position compared to the theoretical position. This curve represents the oscillating (or AC) component of the actual rotating position of the crankshaft, and it is useful in establishing the amplitude of rotational engine vibrations.

An extension of this technique is to measure the same signal at a number of points (a common application is at the front and rear of the engine) and then subtract the deviation curves. This procedure returns the oscillating, torsional component or angle between the measured positions, allowing a measurement of torsional vibrating components across a rotating shaft (like the crankshaft). Once these data are measured, the software may support postprocessing of the data with appropriate display objects for this measurement application, typically, order plots and Campbell diagrams.

There are many stand-alone measurement systems available specifically for this measurement task. However, the advantage of implementing this measurement using a combustion measurement system is that the rotation or torsion data can be fully correlated with the combustion firing events. It is clear that in a combustion engine, the cylinder firing pulses that accelerate the crankshaft are the main excitation force that will initiate or excite any components of mechanical resonance of engine rotating or of reciprocating or static components. Figure 6.5 displays a sample screen showing the torsion analysis feature.

Figure 6.5 Sample screen showing torsion analysis feature.
(Source: AVL Indicom.)

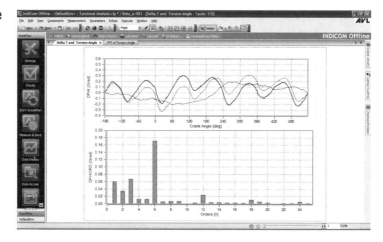

6.2.3 Other Applications

Manufacturers are continuously developing their combustion measurement systems with additional features in the software (and hardware) to support extended measurement functions. This development is generally driven by customer-user feedback, and by the desire to extend the system's functionality so that it becomes a more widely employed and utilised device, rather than a very specialised piece of equipment for occasional use. The modern combustion analysis system is a sophisticated, multiapplication high-speed data acquisition and analysis device. As such, it should be used as often as possible by the owner to recover the purchase price and justify the capital expenditure. Some additional functions and tasks that are available on modern digital measurement system are listed here:

1. *Fuel injection analysis*—A calculation extension applied to standard diesel pump-line-nozzle fuel injection hydraulic systems to calculate the rate of fuel injection. Required inputs are dynamic pressure measured at each end of the hydraulic line, plus needle lift at the injector.

2. *Advanced thermodynamic analysis*—This feature extends the calculation power of the system beyond that of the standard fast heat release calculations that are implemented during the measurement. It provides an environment to execute extended energy balance calculations that consider many more engine parameters and that can hence return a very accurate calculation of the thermal energy conversion and gas exchange process in the engine. It is also possible with certain systems to include a simulation calculation kernel that can return data online that is typically available only from engine simulation code (which is normally executed well after the measurement has taken place, in a postprocessing environment). Combining simulation and measurement at the engine test bed allows immediate establishment of calculated parameters in the test environment, such that these data can be utilised much sooner within the engine development process, hence accelerating that process.

3. *Optical and flame analysis*—Optical combustion analysis can be combined with combustion pressure measurements for a more detailed understanding of flame propagation and temperature. A commonly employed technique is the two-colour

method for diesel flame temperature analysis. In this application, photodiodes measure two specific light wave lengths. The intensities are converted to voltage signals and recorded by the combustion measurement system. Specific features in the user interface of the combustion measurement system support the analysis of flame radiation signals. In addition, this information can be combined and correlated with information from the combustion pressure data to gain further and deeper understanding of the combustion process as it executes in an operating engine cylinder. Particularly useful is the possibility of correlation between heat release and flame intensity. For a premixed flame in a spark ignition engine, the flame intensity and rate of heat release show a similar trend, that is, they are more or less synchronous. Where fuel preparation and mixing are compromised, a diffusion flame occurs, causing delayed combustion with reduced combustion efficiency. Under these conditions, the flame intensity and rate of heat release curves become unsynchronised. It is clear that the combustion process in the cylinder creates flame radiation properties that are a function of the mixture preparation status and quality. This yields quality information that can be appended with the pressure data for a comprehensive analysis and understanding of the in-cylinder processes.

4. *Pulse frequency analysis*—The digital input ports of the data acquisition system can be utilised for further acquisition and analysis of digital signals, in addition to bit change of state and time difference analysis, measured on a crank angle basis. It is also possible to analyse modulated signals. A typical application could be the analysis of a driver signal for a stepper motor engine idle speed actuator. Comparison of pulse duration as a ratio of cycle duration, via analysis of signal edges, can provide duty cycle times.

6.3 Software Interfaces

6.3.1 Remote System

Interface to the combustion measurement system is necessary where the system forms part of the complete engine test bed measuring suite of equipment. The basic requirement is generally to transfer calculated results from the combustion measurement system to an external recording system via a digital interface. This function is analogous to that provided by the analogue output feature. The advantage is that a single digital interface connection, once parameterised, can transfer many results, in digital number form. Hence, errors due to scaling factors, etc., are avoided. In addition, measuring channel inputs and outputs are not required; hence effort and costs due to additional cabling and hardware channel utilisation between the systems to be interfaced is avoided. In addition, this interface can be used to control the combustion measurement system. Often, depending upon the protocol and available commands, the system can be remotely controlled to start and stop measurements, save raw data files, transfer information and results between measuring systems, and load parameter files.

The disadvantage of such interfaces is that, generally, the data transfer speed is relatively slow, which limits the speed of transferring result data to the test bed during the measurement. This can prevent the use of such an interface where deterministic response of an engine control system is required, based on data provided by the combustion measuring system. For example, it may be the case that the engine is to be operated in a control mode that requires engine speed and IMEP control. Real-time transfer of IMEP results during the measurement to the controller would not generally be possible via a remote interface.

For earlier combustion measurement systems, a commonly used interface is RS232. This interface can fulfil all requirements for remote control and data transfer, although the speed of transfer of results during a running measurement is slow, depending on the number of results, typically from 0.5 to 1Hz. This is sufficient for engine steady state measurements where the data will be integrated over the measurement time to produce a single log point result. A more common technique to transfer results for steady state logs is that the combustion measurement system produces statistical results immediately after the measurement, and these results are transferred and stored via the remote interface. This is a data reduction technique that significantly reduces the amount of data to be transferred, yet still allows useful information about the combustion process to be available for analysis. Typical values for each result would be maximum, minimum, average, standard deviation, and coefficient of variance. These statistics provide high-quality information about the results gathered over a number of cycles, but with a much reduced bandwidth requirement for the data transfer connection.

As technology has developed, faster interfaces are now available using standard programming technology (for example, active-X or distributed component object model). This technology allows greater flexibility in programming and configuration of the interface, such that the user can create and write his or her own interface protocol on the host control system. The only knowledge required is the command structure of this interface on the server side (i.e., the combustion measurement system). Implementation using this programming technology is often independent of the hardware layer of the interface, and hence various technologies can be employed. Most commonly used is TCPIP, which is the standard networking technology for personal computers and is widely employed and understood. Using this technology, one can achieve faster transfer of data during the measurement. Typically a 10 to 20Hz update frequency is possible, depending on the number of results. This allows fast screen refresh on the host system for visual observation of the changing trend of a result value, but, more importantly, this speed of data transfer allows the use of calculated results in certain control loops at the host system engine test bed controller. It should be noted, though, that if this mode of operation is used, the results transferred and used in the control loop (e.g., speed versus IMEP) must be calculated via the combustion measurement hardware real-time system to ensure minimal latency. Results calculated or processed on the user interface operating system (e.g., formulas running in scripts) will not be executed in a deterministic way, and they would therefore cause additional delay time in the control loop.

Most combustion measurement systems have similar dialogues to that shown to set up the host and server connection. It is important to correctly configure to determine whether results should be continuously transferred (online, during monitoring or measuring), or whether statistics of the results should be transferred after a measurement over a number of cycles has been completed (or both). Also, correct configuration of channel names is important, particularly when one is connecting systems from different manufacturers, as, in many cases, each manufacturer has its own convention for naming of channels and results.

Typically, the channel of the combustion measurement system (server) is shown as the data set, and the result name expected by the host control system is known as the channel name. Note also that some systems may not support long names, or may require a certain convention to be observed in relation to cylinder numbering or statistical suffixes. This issue could be encountered either on the host or server side, or both.

In addition to transferring calculated results and statistics derived from the raw measured data up to the host data model (i.e., engine test bed automation system) and also remote control of the server (i.e., the combustion measurement system) by the host, it may be required to transfer parameters, typically scalars or text strings, from the host to the server, either before or after a measurement. Typically, this could be where information is required from the host that must be stored with the measured data on the server side, transferred after a measurement. For example, engine information (pressures or temperatures) or descriptive text strings.

Alternatively, this could be data required prior to a measurement, for example, a file name for correlation of the combustion data with the test bed log data. These features are a more advanced requirement, and they are not always supported by all systems (refer to specific manufacturer information).

Error state information is generally also transferable via the remote link. These data are processed and monitored by the host so that any failure of the combustion measurement system can be monitored and logged by the host control system for appropriate action to be taken automatically, or via operator intervention.

6.3.2 Interface to the Engine Electronics Systems

Interfaces among engine electronics calibration systems, test systems, and measuring instruments have evolved and matured such that protocols and interfaces are becoming close to standardised among a large number of motor manufacturers, tier one suppliers, and manufacturers of measurement and calibration systems in the automotive industry. This process of standardisation has been promoted by influential industry bodies, for example, SAE (the Society of Automotive Engineers—www.sae.org), ASAM (the Association for Standardisation of Automation and Measuring Systems—www.asam.net) and AUTOSAR (AUTomotive Open System ARchitecture—www.autosar.org). Thus the possibility of connecting measuring systems to each other, and to vehicle electronic systems, is a reality, as well as a necessity, for current and future development tasks.

A concept promoted by ASAM allows subsystems and modules to be combined in any way necessary for the application and to be integrated via standard interfaces in the respective measurement and test instrumentation environment (e.g., mobile instrumentation, engine test beds). The concept centres around the MCD system made up of the components M (measurement), C (calibration), and D (diagnostics). The model is connected to its environment by three defined ASAM interfaces.

Often, where an engine research and development environment exists at an engine test bed, an environment allowing access to the engine electronics system for calibration and measurement will be available. This environment is used for uploading and downloading of

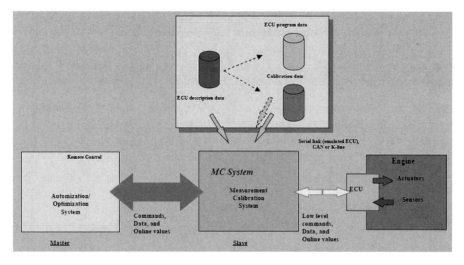

Figure 6.6 Calibration and measurement environment for an engine electronics controller.

engine maps, and for transfer of measurement data during testing. It generally resides on a separate personal computer. It allows direct access to the CPU and memory in the engine electronics controller. These systems also posses a connection for data transfer and control, generally known as an ASAP-3 link. An overview of a typical system is shown in Figure 6.6.

This connection allows access for reading and writing of data via a data server program that can reside on another measurement system. A common example is that this will be located in within a test bed automation system such that ECU data values can be recorded in the automation system, along with values measured via other connected sensors and systems at the engine test bed.

An engine electronics interface can be integrated into the software of the combustion measurement system to allow connection to the measurement and calibration environment. The systems can then communicate via an established protocol, for example, ASAM-MCD-3MC protocol (ASPA3) via TCP/IP, or via an independent serial RS232 interface. The MC system itself communicates with the electronic control unit and receives and writes data to and from engine electronic parameters. The Recorder signals are assigned to engine electronics parameters by parameterisation in the combustion measurement system. The integration of an MC system means that online values from sensor signals or calculation values from the engine electronics can be acquired over time and integrated with the combustion data values, allowing full data correlation. Up to forty online values are normally recorded for each module, typically at 1 to 20Hz per module: for example, for speed, load signal, coolant temperature, intake pressure, etc. Typically, for engine calibration tasks, the interface will be used in reading and writing mode. For integration with combustion measurement data, the task normally involves reading ECU data labels via the combustion measurement system, and storing these data sets alongside the combustion data sets in a single combustion measurement system data file.

6.3.3 CAN (Controller Area Network)

Using the previously discussed ASAP interface is generally possible where a research, development or calibration environment exists because the necessary hardware for ECU development will available, complete with the software environment and interfaces. It is not always possible or practical to have this ECU calibration equipment available during testing, and an alternative method for gaining access to information from the vehicle powertrain control system for integration with combustion data is via the CAN (control area network) bus. CAN is a high-performance bus system originally developed by Bosch for the purpose of allowing vehicle, engine, and power train electronic control units to share information in a efficient way (for example sharing of sensor information). Data could be passed onto the bus and made available to all connected control units (also known as nodes). The main advantage and design feature of the CAN bus is that it can support multiple nodes. This capability allows efficient interconnection of control units via a simple 2 wire bus.

The driving factor that promoted the development of CAN was the increasing complexity of vehicle and engine control systems and the requirement to exchange data among them. This meant that more and more hard-wired signal lines had to be provided. In addition, sensors had to be duplicated if measured parameters were needed by different controllers. Apart from the cost of the wiring looms needed to connect all these components together, the physical size of the wiring looms and the additional weight promoted the development of a system based on a bus network. An additional advantage of this is that it also increases overall reliability due to the reduced number of electrical connections and interfaces.

The CAN bus protocol is highly efficient and secure at transferring data. The data are broadcast on the bus in message format with an identifier. All nodes on the bus receive the message and then decide whether that message is relevant or needed by that control unit. If not relevant, the message is ignored. The identifier also sets the priority of the message so that time-critical information can be processed more quickly than lower priority data. The protocol handles collision on the bus very efficiently (through a process known as arbitration), In situations where two or more nodes attempt to transmit at the same time, a nondestructive arbitration technique guarantees that messages are sent in order of priority and that no messages are lost. This ensures that if a node attempts to transmit on the bus but cannot access it, the data is retransmitted as soon as the bus is free, according to priority. Error handling and message frame integrity are also highly efficient, and that fact allows the CAN bus to operate with a highly secure and redundant philosophy. CAN operates in extremely harsh environments, and the extensive error-checking mechanisms ensure that any transmission errors are detected and managed. The integrity of the system is so well proven that it is now an accepted protocol in other applications outside the automotive industry. It is now widely used in industry as a standard field bus technology for process and production automation. In addition, the high security and integrity have allowed the adoption of CAN technology in the medical instrument industry.

The CAN bus is a serial data bus that can be accessed via a combustion measurement system to allow reading and writing of data to and from the bus. The time-based data in the CAN message frames, consisting of data measured and calculated by the electronic control units, can be read and integrated with combustion data for correlation of engine powertrain control activity with cyclic combustion data. This is done via the combustion measurement system control software, which includes a driver for a CAN interface card normally installed in the user interface PC of the combustion measuring system. Generally, a system such as this can read information from the CAN bus. Each CAN message is transmitted with an identifier that describes the message and its priority. The information is read and displayed in the user interface as a time-based data set. Note that many PC-based acquisition systems cannot be real time because the deterministic performance of the system is not guaranteed due to the windows operating system architecture. Hence, data values from a PC CAN card will not be in real time. In fact, the CAN bus itself, although optimised, cannot be guaranteed as a real-time BUS system. Nevertheless it is still employed in time-critical communication applications on a modern vehicle (e.g., with antilock brake systems and traction control and stability systems).

Note that it is also possible, and could be a requirement, to write value from the combustion system onto the CAN bus. This could be simply for data acquisition purposes, or, it could be necessary where simulation of a nonexistent CAN node is required. For example, it may be that when an engine is installed on a test bed, a vehicle speed CAN message is not available because the vehicle transmission is not installed. Therefore, this node could be simulated by deriving it from another value (e.g., engine speed), then writing that simulated node onto the bus.

Data acquisition via CAN is an increasingly common requirement. Many acquisition systems support CAN and high-speed data transfer via CAN. Collection and storage in a single logging environment is often needed (particularly when testing high-performance engines, either in or out of the vehicle). For this application, quasi real-time performance is needed, and this can be supported by many data acquisition systems, including certain combustion-measuring systems. In this application, the real-time, calculated results have to be available on the CAN bus immediately. In order to support this need, the CAN interface card, which includes the CAN transceiver processor, is installed in the combustion measuring system hardware. Data from the real-time processing system is memory mapped to the CAN transceiver and then written directly to the bus, when received, using an appropriate identifier. Often these CAN messages are multiplexed for maximum efficiency and bandwidth optimisation; this means that each CAN message will transmit up to four results in a single message because the data area of a CAN message is generally 8 bytes. For example, the message may be constructed with the identifier as the cylinder number, then 4 results (maximum cylinder pressure, angle of maximum cylinder pressure, IMEP, angle of 50% energy conversion), each of which is transmitted as a 16-bit number using 2 bytes. In addition, the message contains other standard parts of the message frame, according to the CAN protocol (for example, cyclic redundancy check, start and end of frame identifiers, etc.).

Using CAN in this way allows efficient transfer of combustion results, within CAN messages, for acquisition via a CAN-based data logger that allows comparison and alignment of information from the combustion measuring system with other fast measurement systems. CAN is well-established interface technology, known and used by automotive companies globally.

In general, for combustion measuring systems, the possibility of sharing data between the measurement system and the engine or vehicle electronic control systems, via an established, proprietary interface, is an additional useful software feature. In certain applications, for example, for in-vehicle measurement, this sharing of data becomes a prerequisite for execution of a successful measurement and data acquisition task.

A typical in-vehicle environment would include an ECU data-acquisition system in combination with a combustion-measuring system, collecting data via a single master system (normally the combustion system). In addition to ECU data labels and combustion data, a system for in-vehicle calibration work can often include the possibility to integrate data acquisition modules for measuring temperatures and pressures (typically less dynamic signals) in conjunction with the ECU and combustion data. These modules normally contain the necessary signal conditioning and analogue to digital conversion. Thus, interface is normally direct to a PC or notebook, via a standard interface (Firewire, USB, etc.) This creates a complete data acquisition environment, acquiring signals in a similar environment as would be found at a static engine test bed, thus producing data that can be correlated easily with any test bed data for final engine calibration tuning and sign off, as tested in the target vehicle.

In-vehicle measurement is a developing technology, and portable measuring device are currently being further developed by manufacturers of test equipment. In addition to the above possibilities, test measurement devices for exhaust emissions (gaseous or particulate) are available or under development such that a complete test environment covering all aspects of low and high speed acquisition, as well as special measuring devices is a realistic possibility.

6.4 Calculations and Results

6.4.1 Introduction

One of the main benefits of using a modern combustion data acquisition system for acquiring crank angle based data from the engine cylinders, is the possibility to calculate or extract result data from the raw data, in real time, during the measurement procedure. In most cases the derived results are of more interest and importance to the engine engineer than the raw data itself. An example is the calculation of IMEP from the raw pressure curve. This result is an indicator of the energy or work released in the cylinder and has great relevance and importance, much more than the raw cylinder pressure curve itself. Often the calculated results are used as part of a data reduction technique. They allow the engineer to understand the in-cylinder processes just from a few specific pieces of information. Then, if a deeper understanding is needed, or if more in-depth time-consuming calculations are required, the raw curve data can be referenced and used to facilitate the deeper understanding through an offline data analysis exercise after a series of measurements has been executed.

Many of the important calculated results are key performance indicators for the engine operation and efficiency. As such, they often need to be available in a deterministic way such that decisions can be made by a control system regarding engine optimisation processes, or for engine protection, should an abnormal combustion condition occur that endangers then engine itself.

In the previous sections we have discussed how the combustion measurement system can be interfaced to a control system to transfer test results. In this chapter, though, we will discuss these results with respect to the measurement system, how they are produced and handled, and how they can be managed by the measurement system and the user of that system.

Results can be considered in two main categories. The first category comprises those results that are considered standard, that is, those that are typically delivered with combustion measurement system software and require just activation and parameterisation in the user interface software. Many users, though, will want to know exact details of any preprogrammed, supplied calculations and, if possible, be able to use and adapt those standard methods for their own specific purposes.

The second category is for user-defined results. These calculations may be specific to the user of the system and to the particular application. In many cases, users will have their own methods and algorithms that are protected and that are intellectual property belonging to the system owner. Some facility must be available within the system for the user to program his or her own calculation routines or methods, to be able to execute those calculations during the measurement, and to be able to return the results for immediate viewing or reaction.

6.4.2 Real-Time Results

Real-time results (also known as RTP results—real-time processed results) are those result calculations that are generally delivered with the system. Most often they are hard coded in the firmware of the system processor or coprocessor architecture, and, as such, they can be calculated and delivered by the system to the user very quickly, via the user interface or via some external interface to another measuring system.

The most common system architecture is that the digitised raw data are held in the dynamic memory of the measuring system. This memory is multiported, and the result processor has the ability to read and write data in dynamic memory. During the measurement procedure, as the elements of the crank angle data set vector are written to memory, the result processor accesses the complete crank angle data sets as they are created. Once the data set is fully populated and available (according to the calculation range for a given result), the result processor executes the calculations, generates the results for that cycle, then writes this information back into the dynamic memory for subsequent storage or manipulation. The result processor can be programmed with efficient coding, utilising any optimised instruction sets that are available within the result processor itself (according to type). Thus, the calculated results from the result processor can be available in a deterministic, predictable delivery time. Often the result processor is a stand-alone processing unit dedicated to the task. This setup fully optimises the system and allows a high level of performance in delivering results with full determinism, due to the fact that this processing architecture is handling only tasks for calculation and delivery of results; no other processor tasks or threads utilise the processor resources in a dedicated result processor, so high levels of performance can be achieved with respect to result generation.

Actual levels of performance can be difficult to measure, as factors such as calculation range and resolution, and the complexity of the result calculation all have an effect on the speed of delivery. Certain results need the full crank angle data set over the full cycle range (for example IMEP) so that calculation cannot be executed until all these data are stored in memory, at the end of the cycle. Other results may need only an appropriate operating range to capture the required result, but if the data require a high-resolution measurement (for example 0.1 degree of crank angle is required for accurate timing measurement from a needle lift curve), then there are more elements in the data set vector to apply the calculation to, and thus execution takes longer even if all the required data are available before the end of the cycle. In addition, the number of results set to be returned in the system parameterisation can also have an effect. Clearly more results use more processor resources, and hence loading will be increased and performance reduced, the greater the number of results being returned. For this reason, an optimised setup of the real-time system with respect to calculation range, resolution, and necessity is an important consideration for optimising the system setup for maximum performance.

Generally, a calculation has a number of parameters that must be set correctly and appropriately. For a simple algorithm such as *maximum,* range and resolution are the only parameters. For a more complex calculation, for example, for heat release, further properties must be correctly defined (polytropic exponent, limits, smoothing, etc.). Often, for system standard results, the calculation method is not adjustable; only the parameters of the calculation are adjustable. The reason for this is that the preprogrammed methods are optimised for fast execution; the system user does not have access to the embedded firmware where these real-time calculations exist, as they are programmed into the system source code software build. This arrangement can be limiting for the user of the system. It is important to read any documentation supplied with the system to ensure that the calculation algorithm of interest is fully understood and accepted.

With certain systems an element of flexibility can be added via the possibility of further calculations, and with mathematical operators available to apply to real-time results. This is normally done via a simple dialogue in the software user interface; the execution of these calculations can be done on the personal computer user interface hardware, as they are generally simple and can be executed and made available quickly. This procedure is often known as an arithmetic function, and a working example would be where a calculation may be needed to subtract one result from another (for example, subtract end-of-burn angle from start-of-burn angle to determine burn duration), or to generate statistical information of a result value (e.g., variance of a result over cycles).

6.4.3 User-Defined Results

An appropriate level of added flexibility for result calculation is available in many combustion measurement systems. This allows system users to write and execute their own specific algorithms for display during the measurement, facilitating the possibility of reaction by a control system based on the output of a calculated result.

The most common method of implementation is to allow the system user a programming interface within the system software. This means the user can program his or her algorithm and compile it for execution during measurement run time. This procedure allows freedom for the user to execute his or her own methodology and intelligence during test procedures. Experience shows that every engineer has his or her own idea regarding a calculation method, and therefore requires the ability to execute that in a commercial combustion measurement system environment. Note, though, that permitting this requires that the engineer have a certain prerequisite level of knowledge in programming, even at a very high level within the software (i.e., not that sophisticated). Many engineers will be experienced in writing high-level code for calculation programs, however, not all have this knowledge. Generally, this level of programming is often found at engine test beds, as it is similar to test run programming. It should be noted, however, that if this feature is to be useable, the prerequisite knowledge must be in place.

It is notable that programming interfaces often employ a language structure similar to commercially available programming languages (e.g., visual basic) and that they are relatively easy to learn and use.

The calculations are executed on the user interface system processor (i.e., the PC). This allows flexibility, but it should be noted that calculations running on a PC are not real-time calculations. At best, they are executed online. This means that during the measurement, data are transferred to the PC, and the calculation is executed for display, depending on the system and the interface between the PC and the combustion measurement hardware. This calculation appears instantaneously, and it is displayed on the screen with no apparent delay. Note, though, that the speed of this display depends on the sophistication of the calculation, how many calculations are being executed, and the available resources on the PC CPU. Hence the determinism of the system performance with respect to these calculations cannot really be measured or even stated, and availability of resources could affect the application of these calculations, particularly where they may be needed in engine control loops or for engine protection. Another factor to consider is that, if the calculation requires a crank angle data set as the data source, then often, only one active curve is available during a measurement. It is only after the measurement has stopped that the raw curve data for all the cycles is transferred to the PC. The implication is that if a single result is derived from a curve, then during a measurement, only the result from the current cycle (as displayed) is available. Therefore no rolling statistical information can be derived until after the measurement is completed, and all data are transferred. A working example is that the system user may write and execute an algorithm for engine knock that returns an intensity value. If the mean value of that result over 20 cycles were needed online (for stability), that could not be accomplished via a PC-based calculated formula, in this environement.

Also note that any results calculated on the PC hardware will not be deterministic when transferred via an external interface (e.g., CAN, analogue output), and this fact must be considered where external control loops requiring combustion results are connected to these interfaces.

Nevertheless, in most cases, online calculation and display has sufficient speed and performance to fulfil many steady-state and transient test requirements. The flexibility that the user has to implement and return his or her own methods during the measurement procedure brings a useful extra dimension to the possible applications for the combustion measurement system. Figure 6.7 shows a user-defined calculation dialogue.

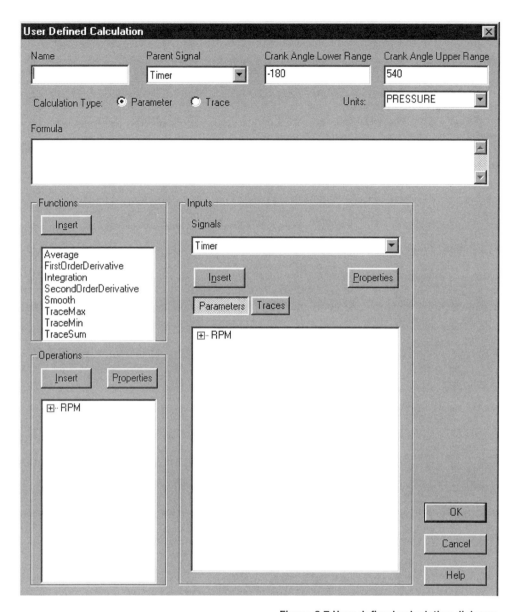

Figure 6.7 User defined calculation dialogue.
(Source: FEV CAS.)

Many companies that offer measurement and test hardware and software have developed graphical user interfaces to assist the user in configuring inputs, outputs, and calculations. Many have adopted drag-and-drop interfaces to allow an inexperienced user to configure a calculation or setup in a logical, graphical way, without having any knowledge or experience of programming. In addition, where complex algorithms or methods are applied, a visual representation can be an effective way of understanding the complex interaction of elements in a sophisticated algorithm. This is a useful feature for even the most competent scientist or engineer. Similar technology is used by many software companies where control algorithm development is undertaken; visualisation of the control model via a network of graphical elements, connected together, allows rapid development and adjustment of a control system philosophy. Then, once finalised, the model can be complied into source code and executed during a run-time procedure. Examples are Labview™ from National Instruments and Simulink™ from Mathworks.

This methodology can be applied to user-defined calculations for combustion measurements. A particularly innovative approach to this problem is the so-called *Calc-Graf* interface from the company AVL. This is a software extension for the combustion measurement operation and data processing interface (known as *Indicom*). The tool allows the user to configure calculations via a drag-and-drop interface. When the operator is configuring a calculation method, predefined calculation blocks are dragged and dropped onto a worksheet and then connected to inputs (i.e., data channels) and outputs (formula channels) via virtual wiring and connectors. An example model is shown below in Figure 6.8.

Once the calculation method is complete, the method can be complied, and the necessary formulas will be created in the background and executed during measurement run time or data post-processing. This procedure provides the user with virtual, calculated channels for display. The contents of the calculation worksheet can be saved for future use (this is known as a model).

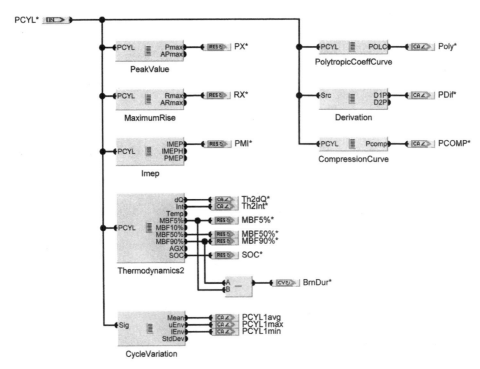

Figure 6.8 Calc-Graf model for further calculations on the raw data.
(Source: AVL.)

The interface provides standard mathematical operators from which a calculation can be constructed, but, in addition, it provides a library of specific algorithms that supports most of the more sophisticated functions that could be required for combustion measurement. Some examples of such a function would be heat release calculations (using various methods), digital filters, knock analysis methods, curve extraction, and reduction, etc. Most usefully, the calculations themselves are not encrypted; the calculation element blocks are known as macros, and, if necessary, they can be copied, adapted, or extended as required. The user can view the detail of the calculation algorithm via a scripting interface or text editor, and the system has a completely open architecture. The user can create his or her own calculations and methods, as function blocks, to be available for use in the graphical environment as drag-and-drop elements. This functionality allows the flexibility to implement complex methods, but allows simple execution.

Another notable feature is the built-in function to support multiple channels. If a multicylinder measurement is made, it is likely that a calculation should be executed on all cylinder channels (example: IMEP would be calculated for all cylinders giving IMEP1, IMEP2, etc.). In addition, it is often the case that result statistics may be required for each result. The Calc-Graf tool handles all these additional requirements that need extra formulas (one for each) automatically via a user selection option in the interface. This dramatically reduces the construction time of the calculation method and model where multiple formulas elements are needed.

6.4.5 Future Developments

It is clear from the the foregoing information that the two current and commonly available mechanisms for handling result calculation from raw data have their own advantages and disadvantages. Real-time calculations can be performed with determinism, but they are less flexible for a user to be able adapt or implement his or her own methods. Formulas executed on the computer interface are much more flexible, but their performance cannot be guaranteed. The next logical step is to implement the facility to program a real-time, deterministic processor, within the combustion measurement system, with the required calculation as defined by the user of the system. This technology is already available in simulation and hardware-in-the-loop test environments and in systems where high-performance processor boards execute user-defined control algorithms with real-time performance. The problem with such systems is that they are laboratory instruments more than test bed equipment. As such, these systems are less physically durable. In addition, they are relatively expensive.

Currently, manufacturers of combustion equipment (that is generally already designed to be durable for in-vehicle or test bed use) are implementing additional processor boards within these units that are externally programmable via the user interface. This provision provides a high level of calculation performance for fast execution of user-specific calculations. Parameterisation of the processor board would generally be via a graphical or script interface, where the algorithm will be written by the user. The algorithm is then debugged and compiled, so that it can be uploaded to the additional processor for execution in real time. Of course performance is still dependant on the complexity of the calculation, but it is likely to be much faster than the PC-based calculation.

This facility will become increasingly significant with the current rate of development in engine and combustion system technology. Engine control units for modern engines have control loops to be optimised. These require combustion-related data immediately, often before the end of the cycle, so that control settings can be adjusted in time for the next engine operating cycle. An example of this needs would be engines running in HCCI (homogeneous charge compression ignition) or CAI (controlled auto ignition) mode. These are the areas where a high-performance, flexible, real-time system is needed. These evolving applications are driving technology developments in combustion measuring systems.

6.5 Postprocessing and Data Management

6.5.1 Introduction

The main purpose of the combustion measuring system is to produce accurate data that relates to the in-cylinder processes. This allows the engineer to gain a clear understanding of the complex interactions that take place as energy is converted in the cylinder to produce work at the crankshaft. Hence, the data, raw and derived, are the main output from the system. Efficient processing and handling of the data are essential to support an efficient evaluation process.

In order to facilitate this, a logical and well-organised data model should form an important part of a fully optimised system architecture. The speed with which the engineer can access data, and the inherent logic of the data model must facilitate a rapid understanding of the quality, plausibility, and meaning of the data. This is a prerequisite to an efficient system architecture.

Data management includes parameter data. This information is essential in order that the system setup be fully understood and traceable when performing data processing. It is important that the engineer fully understand how the system was configured to ensure that any data captured is plausible. The basic settings of TDC, zero-level correction, and parameters used in calculations (e.g., polytropics) must be easily accessible in that parameterisation file structure.

Postprocessing generally involves reviewing measured and calculated data, but, if required, data recalculation can be available via processing software, often supplied with the equipment. This allows users to execute further calculations offline for a more detailed understanding and analysis, often by implementing their own methods or calculations.

Another common requirement for data processing is the exporting of data for use in external programs. The most common format is column oriented, delimited, or separated data. This format is easily imported and read by most processing and calculation programs.

Optimised methods are essential for handling all data related to the combustion measurement for an efficient measuring process, and a logical data philosophy must be integrated, as standard, into the scope of supply of any combustion measuring system and its environment.

6.5.2 Basic Requirements for Data Format and Export

The data produced and required by the system falls in several categories. The primary one is measured and calculated data. Most systems generate this data in a binary format, specific to their system. Hence the supplied software must be used for routine data processing, plausibility checking, and export. The basic data sets are crank angle data, normally split into individual cycles, and cyclic data sets of results derived from the raw data. These are normally fully correlated in the user interface so that raw curves can be plotted in a diagram with the associated result values viewed as tabular or curve data. Once those data can be viewed in this way, further display elements are often offered in the user interface in order to promote a more sophisticated analysis. Often, these display elements are OLE (object linked and embedded) objects, and, as such, they can be cut and pasted into external programs (for reporting and display purposes: for example, Microsoft Word, Powerpoint, or other desktop applications.

It is almost universally adopted that the combustion system software supports the export of data in ASCII format (American standard code for information exchange). This allows the user to export data in a universally recognisable form, to an alternative program for processing (often spreadsheet programs are used in the first instance). This approach is workable, but it is quite time consuming to repeatedly import data columns and format them for graphical display and further calculation. In addition, cyclic data are particularly difficult to handle in this way. If the full, raw crank angle data set is exported, this creates a very large file that can be difficult for the PC to handle in its memory, alongside the demands of the PC operating system itself. Also, some additional scripting or macro work is needed to split the data into individual cycles for analysis. This can be particularly difficult to execute, and the resulting data, from each separated cycle, can be impossible to process with respect to additional calculation tasks. The most common application of exported, column-oriented ASCII data is to use this format for exporting of averaged or envelope curves. This is statistically reduced data derived from the crank angle based raw data. It allows the analysis of a measurement with the requirement to handle all the measured cycles of raw data. Just a single representative curve is produced, and this can be exported easily, as the data set vectors are not too large.

In addition, result tables derived from all the calculated results can be exported as a column-oriented dataset showing cycle number as the abscissa versus all calculated results. This is again a reduced but valid data set for the measurement that provides the engineer with the required information from the measurement task. In general, handling the raw, crank-angle based cyclic data should be limited to analysis in the measurement system native data format, via the supplied software for reading that data format. This approach allows detailed viewing of the raw curves on a cycle-by-cycle basis if required. In addition, the combustion measurement system–supplied software will handle the complexities of crank angle datasets with different ranges and resolutions in the same display windows. This is a difficult task for an external program that is not specifically designed for raw crank angle data. From this software, reduced data sets can be exported for processing or display in external programs via ASCII data formats.

Note, though, that an additional benefit of using the combustion system supplied software for processing (often known as an offline licence) is the automatic correlation of the raw curves with a respective engine volume table curve or dataset. This is the most significant factor that allows further offline processing and analysis. The volume table is derived from the engine parameters and the crank angle position and when available, allows further re-calculation of work and energy using different methods and philosophies. Where curve data is exported to ASCII, this engine parameter information can be held in a separate file or location. When the information is stored in that way, it is not obviously linked or connected to the raw curves that it belongs to. This means that recalculation of the volume table in the processing software is required, as long as the engine parameters to calculate volume and the crank position for each pressure measurement point are known.

Another option for data export is to read the native data files from the combustion measurement system. These are most often stored in binary file formats, optimised for fast reading. The implication of this is that a software interface must be written or created in order to be able to read the data correctly. Many engineering and scientific analysis programs incorporate flexible programming interfaces, and as long as the file description and format are available for the data source, a programmatic interface can be written to read in the data bytes for graphical display and calculation. Another option is that data can be read via a plug-in interface to the software, typically a DLL (dynamically linked library). Many commercial programs provide this level of interface, and using this technique, although more complex to implement, gives better performance when one is reading the data into the PC memory for processing.

6.5.3 Requirements for Engine and System Parameters

The parameter files for setup of the measurement task are as important as the data itself. The information included in the parameter file includes the following:

- Engine geometry and information
- Channel and signal definitions
- Display object configuration
- Measurement system setup

This provides the basic structure of the system configuration for a task. Generally if the task or application is changed, the complete parameter file is exchanged. In some cases the information is held in a single file that includes all this required information. Certain manufacturers of equipment also separate out files for test definition, signal definition, and screen display definition. These files are generally contained and handled in a single, project-related file by the user. This allows a certain level of flexibility when one is configuring the system. For example, if the engine changes, TDC setting will need to change, but signal definitions and screen displays will remain the same if the test application is unchanged.

An alternative approach is a complete project-based work file that contains all required information as just stated. In addition, the file can contain calculation methods, formula, scripts, and display objects so that all the information required for a complete measurement and analysis task can be managed in a single file. This is an intelligent approach as measurement applications become more sophisticated. The integration of scripts to execute file loading and processing tasks automates processes that take time and require a certain level of expertise that may not always be present.

Additional components that can be found in a project-based method for handling task and measurement parameters are listed here:

Formulae and calculation methods

Scripts for automated data loading and manipulation

Additional display objects integrated in windows (specific bitmaps)

In addition, there are certain parameters relating to the measurement system that are not routinely changed but that must be accessible to the user, either via the user interface, or via a text editor. These are generally system configuration settings, for example, activating additional inputs or outputs, or setting calculation or performance limits. Generally, this parameter information is accessed and stored as an ini file. If a change is made to the settings in this file, then normally, the measurement system must be rebooted to activate the new setting. This setting in this file should not be adjusted unless the user is completely familiar with the system and the consequences of the change. Adjustment of these parameters should therefore be considered service adjustments.

Another consideration with respect to the system setup relates to configuration presets, a typical example being default data paths or files names. These are parameters or preferences that must be set by the user, but that are often not contained in the system parameter file or the measurement data file. The system user must set the commonly utilised data paths for storage such that the files are sent to the directory where they are expected, either locally or at a host location. In addition, it is normally the user's responsibility to set up a logical folder structure that supports measurement and parameter data for a specific measuring task or project.

File-naming convention is another important factor, particularly if a series of measurements is performed at different engine operating conditions. Often it is required that a default file name be employed with incremented file extensions, or a specific file name structure might even may be required. In either case, the main reason for these procedures is to prevent files from being overwritten where multiple measurements are made in a sequence. Often all measurement data and parameter file would be saved in a single file folder, with subfolders for parameters and data. Disciplined file handing is important to prevent the loss of data, or a lack of understanding of where the data has come from!

6.5.4 Typical Environment

In practical terms, the most common requirements are to be able to execute a measurement task, employing a certain system setup, and to be able to save all the related data in a single, efficient file format. In addition, to being able identify the file during postprocessing in such a way that it is easily traceable and retrievable, constructing logical links among all these tasks with respect to the files involved is an important factor to achieve and efficient work flow processes.

The system setup information is useful where unexpected trends or results are seen in the data. This environment allows the user to trace back and examine the system configuration to ensure that all relevant parameters were correctly entered. This is particularly important with respect to engine information or channel calibration settings. This information is generally held in the parameter file, although some basic engine geometric information could also be stored in the measurement data file. Certain systems can also support integration of the parameter file with the measurement data during a data save operation. This is a very useful function, but it increases data saving time significantly. A most useful feature is the possibility of having the parameter file name, embedded in the data file (as a scalar operating parameter). This arrangement allows the required traceability, assuming the parameter file with that name has not been changed since the measurement was taken.

Generally the user can use a default file name and incremented extension, or each measurement data file can be named via a user interface dialogue while one is saving the data. For automatic operation of the combustion system in conjunction with the automation system, automatic file name handling is essential. In addition, the host control system will often pass additional measured engine-related values into the combustion data file for reference (e.g., writing comments lines) or for processing (e.g., measured values of interest—manifold pressure, temperatures, etc.).

Where the integration between the test automation system and the combustion measurement system exists, typically, bidirectional data flow occurs, but usually the most important function is to pass combustion results into the test bed data model. This is a typical, steady state measurement requirement that allows a full set of test bed and combustion results at each engine measurement point. In cases where these raw data are saved, the file name used, as dictated by the host controller (test bed automation system), is stored at each log point in the test bed data model. Therefore, when one is processing data, all the combustion results are available at the log point (statistically reduced, for example maximum cylinder pressure values—maximum, minimum, average, and coefficient of variance), and it is possible to trace back to the raw data file, to view individual cycles, or results from those cycles, if a measured result is identified with a concern. In many cases the raw data are not saved at all on the combustion system. This arrangement is typical where a fast interface exists that passes calculated results from the combustion measuring system to the host test bed automation system. This could be an analogue output system, or, alternatively, a digital interface (TCPIP or CAN). With this method, combustion results are continuously polled and transferred to the host via this interface for screen display , or for saving data via a recorder measurement (transient application) or log measurement (steady state—integrated value). In this case, time delay due to latency of the interface must be known and accepted. Analogue interfaces are generally real time, and they are measured as an analogue channel with no delay. The limiting factor here is generally the number of available analogue output channels. In most cases the maximum number is sixteen (i.e., for a four-cylinder engine, four results per cylinder). Digital interfaces are generally not real time, but they can have sufficient performance for certain applications (i.e., calculated result to be used in a test bed control mode—speed versus IMEP.

Chapter 7
Applications

7.1 Introduction

In this section of the book we will look at specific areas of the combustion measuring chain and focus on the aspects that really differentiate a combustion measurement system from other measurement devices found in a typical engine test cell. The so-called *measurement chain* is the collection of components that together form the combustion measurement system. One doesn't need to be an instrumentation expert or a scientist to use a combustion measurement system. The user of the complete system needs a good understanding of each main component, its function and operation, so that when problems occur, either during or after measurement, the system user has sufficient knowledge of system operation, but, more importantly, of the application, so that progress can be made.

An essential factor is to understand the scope and target of the basic measuring task, and of the fundamental characteristics and operation of the unit being tested (i.e., in general, a combustion engine). Assuming that this fundamental knowledge is clear, the user then needs an understanding of the specific aspects and technical details of the measurement procedure, the analysis, and the technicalities of the combustion measurement in order to produce data successfully and efficiently.

This section of the book focuses on the latter part, and it explains the specific, idiosyncratic details of combustion measurement with respect to the engine operation and measurement task, to facilitate the required level of user understanding.

7.2 Measurement Chain Properties

7.2.1 Introduction and Overview

In the previous chapter we discussed the main components of the measuring chain, and how their technology is utilized individually to measure combustion. We also identified specific characteristics of the components with respect to the application. The most important consideration is how these subsystems connect and work together, and how the characteristic features of the components are handled in a combustion measurement system. That is, we answer the question, what are the specific features in the combustion measurement device that make it what it is, as opposed to being just a high-speed data logger? Here is a review of the main components of a combustion measurement system:

1. The sensor and adapter mounted at some position on the engine, according to the task requirements, with appropriate technology employed in the sensor.

2. The signal conditioning system to convert the raw signal for the sensor into an optimised, usable signal for the measurement system. (The associated cabling interfaces are part of the signal conditioning system.)

3. The angle encoder (or other device) to create an incremental pulse in relation to crank rotation as the crank rotates, plus associated cabling.

4. The measurement system itself, including hardware, software, and any external communication interfaces.

Specific technologies used in the system are less commonly found in general measurement applications, particularly in an engine test bed environment. For example, the encoder system for a combustion measurement is far more complex than a normal encoder used for simple speed measurement (which is often just a sensor and trigger wheel). The encoder has to be able to withstand engine vibration, yet accurately produce marks for crank- and cycle-based information. This is just one example; in this section we will discuss others.

7.2.2 Special Considerations for Combustion Measurement Instrumentation

There are a number of main points to consider with respect to combustion measurements.

Measurement in the Angle Domain

Most other engine test bed equipment and external measurement devices are focussed on two main measurement modes: a steady-state, integrated measurement producing a single log point result value, or a transient measurement that gives a continuous changing value over time. The combustion measurement system measures in the angle domain, and thus the data produced have completely different abscissa when compared to test bed data. In addition, the data are acquired at much higher frequency, but not at a fixed frequency, and the acquisition frequency depends upon engine speed. Of course engine speed is a fundamental aspect to be considered.

Angle Encoders

As already mentioned, the angle encoder used for combustion measurement is quite specific to its purpose. It has to be located and installed carefully on the engine, it has to be designed specifically to withstand the heat and vibration levels encountered, and it has to reliably produce incremental angle degree and revolution marks. In addition, the cabling and signal transfer must be effected reliably with minimal possibility of electrical noise pickup. The angle encoder signal is entirely responsible for triggering the digital sampling of the measured channels. It is important that the encoder be correctly set up to generate the angle marks at the correct resolution for the measurement task in hand. The encoder system (in conjunction with the measurement system) must be capable of producing marks continuously and reliably at the highest measurement resolution required, right up the anticipated maximum engine speed.

Pressure Measurement

The piezoelectric measurement chain needs quite specific knowledge and handling. The transducers are very small and delicate, and great attention to detail is required during the adaptation process when instrumenting the engine. One must exercise precision and care when installing and removing the transducers from the engine, and one must pay particular attention to the manufacturer's stated parameters for installation (location of the transducer, installation torque, etc.). The transducers themselves must be kept clean to laboratory standards, as must all the measurement cabling and connectors.

A most important aspect for combustion chamber pressure measurements, when compared to normal pressure measurements around the engine test cell (for example manifold pressure, oil pressure, etc.), is the nature of piezoelectric measurement. As discussed previously in this book, the piezoelectric sensor effects a dynamic measurement. It can measure only the change in pressure, not the absolute pressure itself. This means the combustion measurement measuring chain must be able to compensate this weakness of the piezoelectric measurement system and somehow establish and correct the measured curve for atmospheric conditions at the beginning of each engine cycle.

Signal Conditioning

The piezoelectric measurement technique is a reliable and robust technique, with sensor characteristics ideally suited for the application. However, there are inherent problems that must be considered in the signal conditioning of the charge-to-voltage converter (charge amplifier). The amplifier itself performs very high-gain amplification, and the input side of the amplifier is very susceptible to problems if it is not kept rigorously clean (including cleanliness of cables and connectors). Any path for current leakage that is created by a general lack of cleanliness (as is generally the case anywhere engines are being operated) will have an adverse affect on the measured data and on the success of the measurement procedure itself. It can be a complex task to properly set up amplifiers of any type. If the amplifier gain setting does not use the full input range of the measurement device, there is a risk of poor-quality digital conversion that can seriously affect any calculations made, based on the raw data. In addition, for some applications there is a risk of losing important information completely from the pressure curve data.

Measurement Hardware

Correct parameterisation of any measurement system is always important, and for combustion measurement it is very important. The reason for this is that the results of interest (i.e., the measured data) are at the very end of a long and complex measuring chain, with many interdependencies among system components—that is, there is a lot that can go wrong. Many errors can be put down to simple mistakes or oversights, though such errors are not always very apparent during measurement. A combustion measurement procedure generates very large data files that cannot always be closely inspected during the test. It is only with operator experience that measurement quality performance indicators (during the test) become visible to the engineer, and often the combustion measurement is carried out by an operator with limited knowledge of the details or with limited experience with the measurement task. This lack of operator knowledge brings with it the risk that, when the needed data are inspected closely, errors will be apparent that reduce data integrity. Even worse, often by this stage the engine has been removed from the test bed, so a retest (costly as it would be) cannot be done, and the test turns out to have been a waste of time and money!

The combustion measurement system is different from most other engine test bed devices in that the number of channels measured, compared to those calculated or derived, is low (most other devices are the opposite). The basis of the task is to measure a curve against crank angle with the correct position in the Y (i.e., measured unit) and X (crank angle) axis. Many calculations and derivations are executed from this simple data set. Therefore, any slight error with respect to this curve and its position relative to the measuring scale and abscissa can cause massive errors in calculated data (at least on the order of a magnitude).

This factor should also be considered by and apparent to all those involved in using the equipment.

External Interfaces for Control and Data Transfer

Very often in modern automated test environments, the control of combustion measurements is handled via the test bed automation and control system. That allows remote-controlled operation of the equipment as a part of the test bed measurement environment. This procedure requires careful setup, in addition to a solid understanding of what is required from the test itself. The parameters for the remote interface are complex, and what their function is might not always seem that clear. Time spent defining the requirements and setting up and testing the operation of the interface before the actual test time can be a worthwhile investment, because such pre-setup and testing will facilitate seamless operation, and consequent harmonisation of the test cell instruments and systems as a holistic measurement environment.

7.3 Zero-level Correction, or Pegging

7.3.1 Introduction

As mentioned previously, in most cases, the combustion pressure measurement consists of a piezoelectric sensor and charge amplifier, whose output interfaces with the combustion measurement device. Also, we have discussed that a piezoelectric sensor can measure only dynamic pressure, not static pressure. The main implication of this is that the measured cylinder pressure curve, direct from the sensor, shows the change in pressure during the engine cycle, but it does not show the absolute pressure. Therefore, for processing and calculations, the curve has to be shifted by a correct offset value so that it represents an absolute, measured curve. This is particularly important because a correctly referenced curve is necessary for accurate establishment of energy release curves. Although, for certain other evaluations and result calculations, the zero level of the curve has no effect (IMEP and position of pressure maximum for example). Also, it is worthwhile to note that zero-level correction on a cyclic basis removes many of the problems associated with long-term drift inherent in a piezoelectric measuring chain.

There are a number of established techniques, well described and proven in technical literature, that present and compare various methods of correction. These methods also vary in their complexity, processing time, and accuracy. Generally, some are more suitable than others, depending on the application, and they all have advantages and disadvantages, so it is up to the person responsible for setting up the equipment to decide which method is most suitable. The most commonly encountered methods are described here.

7.3.2 Fixed Point and Reference Value

The fixed point and reference value method is the simplest method: at a fixed, specific crank angle (i.e., at a reference angle), the whole pressure curve is shifted on the Y axis until, at this point, a predefined pressure value (i.e., a reference pressure) is achieved (that is, until the difference between the measured and reference value is zero). Often the measured curve is filtered or averaged around the reference angle to minimise the effect of any noise or spikes on the signal. Figure 7.1 shows the effect on the curve of this method.

Applications

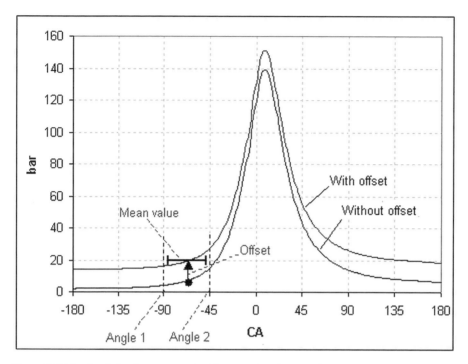

Figure 7.1 Fixed-point reference method for zero-level correction.
(Source: D2T.)

This method is suitable only for unthrottled, unboosted engines, with respect to cylinder pressure measurement, where the reference pressure value is assumed to be atmospheric at the reference angle. That is generally not the case in modern engines, and therefore current applications for this method are limited. Fixed point and reference value testing is, however, suitable for use on other types of measured channels that may need correction on the Y axis for screen display. For example, the method works well for needle lift or ignition signals.

7.3.3 Fixed Point and Measured Value

The fixed point and measured value method is an evolution of the fixed point and reference value method, and it involves measuring a signal to use as a reference value at a fixed crank angle. The curve is shifted every cycle in the Y axis until the measured pressure curve and the measured reference sensor curve have zero deviation at the reference angle. This phenomenon is shown diagrammatically in Figure 7.2.

This method is an improvement over the fixed point and reference value method because it uses a measured value for referencing (the offset channel). This would normally be a single, absolute measuring sensor, measuring the manifold pressure, acquired on a channel of the combustion measurement system. Generally it is possible to average this channel over a number of crank angle data points to remove unwanted effects of noise or spikes on the signal trace and to provide reliable correction of the measured curve on a cyclic basis. For

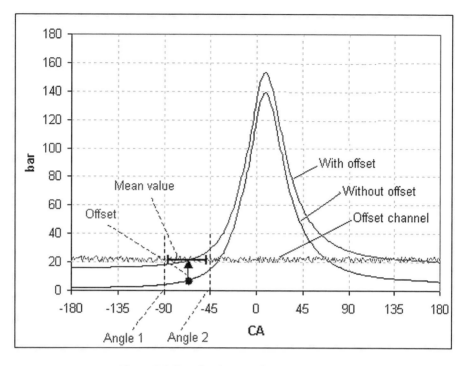

Figure 7.2 Zero-level correction via measured value at fixed position
(Source: D2T.)

extreme accuracy of measurements, a cylinder-specific, absolute measuring sensor could be installed for the instrumented cylinder, as close to the inlet valve as possible. This sensor then measures the actual pressure in the specific cylinder and inlet port at the crank angle where the reference value measurement is taken, as opposed to an average value shared across all cylinders, as is the case when measuring the mean intake pipe pressure via the inlet manifold. If the measured pressure data will be used for detailed subsequent calculation of energy balance, and for the highest level of accuracy when comparing simulated and measured data, this latter method (cylinder-specific sensing) is recommended.

7.3.4 Thermodynamic

The thermodynamic method is applicable only to cylinder pressure signals, and it employs a simple thermodynamic calculation to derive a pressure value, at a reference angle, during the compression phase of the engine cycle. This value can then be used as a reference point to apply an offset on the Y axis and to shift the measured pressure curve, on a cycle by cycle basis. The algorithm assumes that the compression process is a constant polytropic process between two crank angle positions. The pressure difference between these two angles is of course given by the sensor and the measurement. In addition, the volume at each position and the change in volume are also known, because they are derived from the crank degree marks and the engine geometry. If process exponent is known or can be estimated, the pressure value for the offset correction can be calculated from the following simple equation:

X1: first angle value (typically: −100 °CA)

X2: second angle value (typically: −65 °CA)

$$P_1 = \frac{\Delta p}{\left(\frac{V_1}{V_2}\right)^\kappa - 1}$$

P_1 Pressure at CA X1

V_1 Volume at CA X1

V_2 Volume at CA X2

κ Polytropic coefficient

The result of this correction method is shown in Figure 7.3.

This method provides a quick, real-time method of cyclic zero-level correction that, in most cases is accurate. The main problem with this method is that if the chosen angles for the reference points are subject to any noise or spikes on the pressure curve, that noise can seriously affect the calculation of the zero level, causing the pressure curve to shift erratically and completely off the viewing screen (thus invalidating that measured cycle). Typical values for the calculation points that are generally used and recommended are from 100 to approximately 70 degrees crank angle before the top dead centre (TDC), though in this range there is a significant risk of the effects of structure-borne vibrations from intake valve motion. This is a problem, particularly when one is measuring high-performance engines, but the situation can be improved to some extent through experimentation by shifting the upper crank angle position. Note that the reference points must always lie between inlet valve closure and start of combustion.

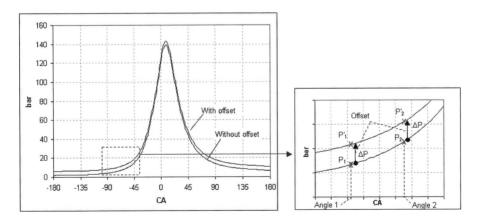

Figure 7.3 Thermodynamic based zero-level correction
(Source: D2T.)

To resolve this problem, some combustion measurement systems have the ability to execute an averaging window on the measured curve around each chosen reference point on the compression curve. The width of this window is selectable by the user, according to the level of noise present on the signal (typically ± 4° of crank angle).

Note also that this method assumes a constant polytropic process between the two defined points on the compression curve. In reality, a constant polytropic process of course does not exist, and this fact can be a source of error. When one is using this method, the distance between the two chosen angles should not be too great, in order to minimise this effect. Typical values for this exponent that describe the compression process are are provided here:

> *Compression ignition and gaseous fuelled engines—1.37 to 1.40*
>
> *Spark ignition (SI) engine with homogeneous charge—1.32 to 1.33*
>
> *Spark ignition engines with homo-heterogeneous charge—approximately 1.35*

These figures should be used as a guideline only. It is easily possible with most combustion measurement systems to actually measure and establish the polytropic index accurately in the required range. This is particularly important for situations where a polytropic index is not known, for example, when one is using very fuel weak mixtures, or when one is developing and using advanced combustion technology concepts.

7.3.5 Alternative Methods via Post Processing

One can establish the correct zero level during data processing by closely examining the data, plus making comparisons with data produced by an engine simulation model. Postmeasurement adjustment provides high accuracy and integrity, and it is usually performed as part of an analysis procedure. The measured curves can be compared with the simulation, and they can then be shifted until the area between the measured curve and simulated curve is at a minimum in the position of the curve of particular interest. This technique provides the correct referencing in the part of the cycle that is most relevant to the measurement target. For example, accurate referencing of induction and exhaust pressures is important for studies of engine breathing and friction measurements. For detailed combustion analysis, it is more important that the high pressure part of the curve be accurately referenced. During this procedure it is also possible to make other adjustments, for example, to the compression ratio or to the crank angle position of the data.

The integrated heat release curve can be examined closely to verify correct positioning of the zero level, particularly during the compression phase, as it can be assumed that no energy will be released during this process. The pressure curve can be shifted iteratively until the integrated heat curve approximates a zero line throughout the compression process.

The advantage of these methods is that they provide highly accurate establishment of zero level and are therefore suited to situations where detailed analysis of the measurement data will be undertaken. The disadvantage is processing time, because these methods cannot be executed in real time during the measurement. The computation time and effort are too great. In addition, the user must be very familiar with the measurement technique and with engine thermodynamics to ensure good quality and reliable data adjustment.

7.3.6 General Comments

Because of the number of methods available for zero level correction (or *pegging*) the engineer or other combustion measurement system user has to make an informed decision as to what method is most appropriate to the task at hand. The thermodynamic and pressure referencing methods are popular and well established for correction in real time during the measurement, though each has its advantages and disadvantages. In many cases, the method selected comes down to personal preference, but other factors that affect the quality of the chosen procedure should also be taken into account:

Transducer effects, i.e., thermo shock

Accelerations and vibrations from the engine components, i.e., valve noise

Other possible interferences, i.e., electrical noise

The table in Figure 7.4 is provided as a general guide for the comparison of the methods discussed.

Assessment of zero-line detection methods

	Fix point adjustment with ambient pressure	Fix point adjustment with mean intake pipe pressure	Fix point adjustment with CA-resolution-based intake pipe pressure	Constant polytropic coefficient	Integral of Heat Release / comparison measured-calculated pressure curve
Peak pressure p_{max}	o	o	+	+	+
Duration of combustion	-	o	o	o	+
Combustion delay	-	o	+	+	+
Position of 50% energy conversion	-	o	+	+	+
Energy balance	-	o	o	o	+
Additional metrological effort	no	(yes)	yes	no	yes
Accuracy	moderate	moderate	good	good	very good
Method suitable for realtime	yes	yes	yes	yes	no

+ ...... suitable
o ...... suitable to a limited extent
- .... not sufficient

Figure 7.4 Comparison of zero-level correction (pegging) methods
(Source: AVL.)

7.4 TDC Measurement
7.4.1 Introduction

For combustion measurements and analysis, it is imperative that the measured cylinder pressure curve be accurately referenced in both the X and Y directions relative to the respective axis. Zero-level correction performs this function with respect to the Y (ordinate) axis. A procedure for TDC (top dead centre) measurement and establishment performs this function in the X (abscissa) axis.

Angle encoders used in combustion measurement are generally incremental, with two signals, one of crank degree marks, and another of pulse per revolution signal. The former is used to identify the position of the engine crankshaft, and the latter (often known as a trigger mark) is needed to establish the absolute position of the crank degree marks. During installation of the encoder, it is not always possible to know, or to determine accurately, the position of this trigger mark relative to the true crank position. Therefore, some method and procedure must be executed to measure and establish the position of the trigger mark correctly with respect to the absolute position of the crankshaft. This procedure is generally known as TDC measurement. Once the TDC offset is established, the measured crank degree marks can be shifted in the software user interface of the combustion measurement system to determine crank position absolutely.

Determination of the crankshaft marks in absolute terms is imperative to the accuracy of subsequently calculated results. The slightest error in the crankshaft position is magnified by at least an order of magnitude with respect to calculations such as IMEP (*indicated mean effective pressure*). Crankshaft position must be determined with respect to TDC position within 0.1 degree to accurately calculate this result. Figure 7.5 shows the impact of incorrect determination, that is, of angular error on specific calculated results.

If the cylinder pressure curve position relative to crank degrees is defined incorrectly, for example, TDC too early, this has the effect that, for any given angle position during compression, the measured curve shows lower than actual values; during expansion it shows higher than actual values. As a result, higher energy conversion values are apparent (see left diagram, Figure 7.5). The opposite effect occurs if TDC is late.

The effect of angular error can also be seen on FMEP (*friction mean effective pressure*) (right diagram, Figure 7.5); when TDC is too early, larger FMEP values are apparent.

The consequence of correct TDC determination cannot be overemphasised. It is very important to successfully acquire the measured curves correctly with respect to crank position. Any shift in the relationship between the pressure and the volume (derived from the crank degrees and the engine geometry) has a significant effect on measurement data quality, making the data mostly invalid. Based on this fact, all combustion measurement systems have various procedures built into the user interface software to guide the user through a successful TDC determination procedure. Safeguards include features to flag or reject suspected incorrect measurements.

Applications

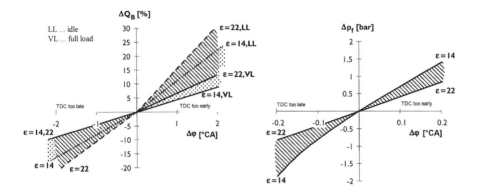

Figure 7.5 The effect of crank position (TDC) error on the energy balance and FMEP values calculated from a diesel engine.
(Source: AVL.)

A number of methodologies in use involve varying levels of effort in return for various levels of accuracy. We can state that, in general, the techniques with lower cost and effort produce lower levels of accuracy. For some applications, the compromise can be acceptable. The main methods are discussed next.

7.4.2 Methods

7.4.2.1 Static Determination

The static determination method involves mechanical and manual intervention at the engine. The static TDC position can be established through measurement of piston displacement, and the flywheel or pulley (or other external accessible rotating part directly connected to the crankshaft) can be marked to show the correct position (using paint or etching).

The procedure requires access to the piston with a dial gauge precision-measurement device. One rotates the crankshaft to an angle of approximately 90 degrees after TDC. The dial gauge is installed via the spark plug or injector installation bore (assuming the cylinder head is in place) to measure the piston height. The reading is noted at this position. In addition, the pulley or flywheel (rotating component) and cover (static component) are marked.

The engine is then manually rotated in the normal direction of rotation, with the effect of slowly lowering, then raising, the piston until the exact same reading is achieved on the dial gauge. The pulley is marked again relative to the static mark.

There are now 2 marks on the pulley and one on the cover. Each pulley mark is exactly the same distance from the TDC position, one before, and one after. Therefore, the centre position between these 2 marks (on the pulley) is the precise static TDC position. This can be measured and marked. If the engine is rotated to this TDC position when the encoder is installed, and the trigger mark is observed and adjusted so that the trigger edge occurs at this point, the trigger is calibrated to be at the static TDC position. Figure 7.6 shows the general arrangement for this procedure.

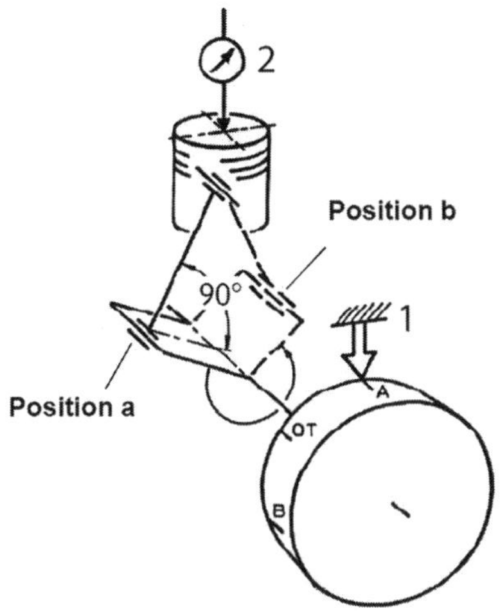

Figure 7.6 Method for static TDC determination
(Source: AVL.)

Note that it is quite common for engine manufacturers to mark the front pulley or flywheel of production engines with a TDC mark to be used for monitoring the ignition timing at routine maintenance intervals.

When one is executing this procedure, it is important that an angle with sufficient distance from TDC is chosen for marking the pulley. The reason for this is that piston movement around TDC is very small per degree of crankshaft movement. The optimum position is around 90 degrees before and after TDC, but the actual angle will depend on the reach of the measuring equipment and the availability of access to the piston crown.

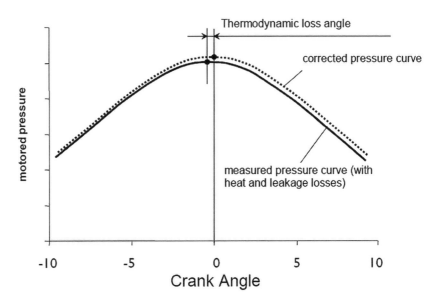

Figure 7.7 Thermodynamic loss angle
(Source: AVL.)

7.4.2.2 Pressure Curve Determination

Pressure curve determination is a very popular method of TDC determination that is supported by all manufacturers of combustion measurement equipment. It is a well-proven technique that allows establishment of the dynamic position of TDC, that is, TDC with the engine rotating, as opposed to the simple static determination described above.

The fundamental principle behind this procedure is to rotate the engine and measure the cylinder pressure curve of the unfired engine, a so-called *motored curve*. From this curve, it can be assumed that the maximum, peak pressure value occurs at TDC, that is, that the smallest cylinder volume exists at TDC. One might therefore assume that the position of the maximum value can be taken as TDC. Actually, this is not the case!

Due to the real-world thermodynamically imperfect compression and expansion process, the measured pressure curve peak actually leads the real TDC position peak by a certain angle (shown in Figure 7.7), which depends on a number of conditions (engine speed, temperature, blowby, etc). This discrepancy is known as the loss angle, and when one is determining the TDC using the pressure curve method, this loss angle must be measured or estimated by the user and input into the system in order to derive the correct TDC value. Figure 7.8 shows loss angles for typical engine types at various engine speeds. Note that loss angle increases at lower engine speeds because of the increased cycle time and, hence, the longer time available for heat and pressure losses.

In practical terms, one can achieve engine motoring very easily if the engine is on a test bed attached to an active dynamometer. If this environment is not available, the engine can be motored with the starter motor or via other fired cylinders via deactivation of target cylinder firing during measurement. The procedure can vary according to the equipment

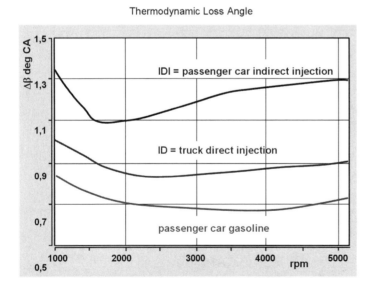

Figure 7.8 Typical thermodynamic loss angles
(Source: AVL.)

manufacturer, but, in general terms, there is a specific calibration procedure built into the combustion measurement device. Once this calibration procedure is selected and activated, the user selects the signal (or cylinder) to be evaluated. The system then prompts the user to start the measuring procedure.

Once the procedure is initiated, the measurement starts, and the appropriate number of cycles is collected. Then the evaluation is executed by the system to return the correctly adjusted TDC value for the specific cylinder, taking into account the loss angle. This value, if accepted by the user, is generally stored in a parameterisation file. The TDC procedure then needs only to be reexecuted to check that the TDC position is correct (for example, to check the integrity of the angle encoder installation), or if the angle encoder has been disconnected from the engine for any reason.

Numerous theories have been proposed for the correct determination of TDC from a motored curve, and these depend greatly on the manufacturer of the equipment. Simply determining the peak position of the curve is not sufficiently accurate for TDC determination procedures. The change in piston movement relative to crankshaft movement is very small around TDC, and therefore the change in pressure is also correspondingly small. That small change in pressure means that the pressure curve is relatively flat around TDC. Thus, in this limited range, digital conversion is least optimal. This fact makes exact determination of the peak more challenging. In most cases, a more sophisticated analysis of the curve must be executed. The most commonly used methods are discussed here.

AVL

This algorithm is executed on a single cycle to determine the correct angular position of the peak of the measured curve, before applying the loss angle for establishment of TDC. The measured, motored curve from the reference cylinder is bisected at equidistant points before and after TDC (-14° to -4°, then 4° to 14° of crank angle, for example). A straight line is connected from each symmetrical point. The centre position is derived from that straight line. This procedure is carried out at all the above ± crank angles. A best-fit straight line is then drawn, interpolated through the multiple centre points in order to derive the centre TDC position of the curve. This method produces a TDC offset value, plus a standard deviation result relating to the individual symmetrical points. The results of this procedure provide an indication of the quality of the curve symmetry and hence the accuracy of the TDC position derived from the curve. This value is also used as a plausibility indicator, because if the standard deviation is greater than 0.3°, the calculated TDC value is not accepted by the system.

FEV

FEV uses a similar method to AVL, except that it is called the interval method. The main difference between the two methods is that the FEV evaluation is carried out on an averaged pressure curve that is the result of a measurement over a number of cycles. Figure 7.9 shows the approach.

An alternative approach from this company is the so-called *polynomial method.* In this procedure, the system measures and calculates an average curve from a number of

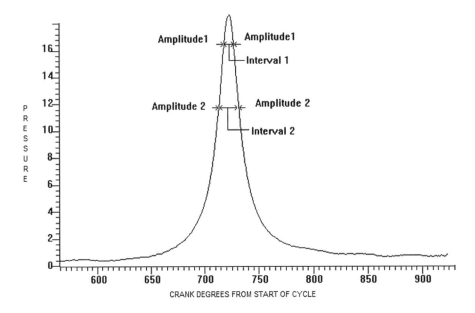

Figure 7.9 Interval method of TDC determination
(Source: FEV.)

consecutively measured cycles. The system then fits a second-order polynomial around the peak of the calculated curve. If the measured values at the peak deviate from the calculated curve by more than 5%, then the measured values at ± 1 degree around the peak are eliminated. This process is repeated iteratively until the deviation is less than 5%. Once this is achieved, the point of inflexion of the curve (i.e., the maximum value) is taken for the TDC value.

This process is a simple curve fitting to establish a peak numerically from a simple, second order curve that has been fitted to the measured data. It is an alternative to the former method that could produce better results where measurement of a smooth, motored curve could be challenging.

D2T

This company offers the same method as AVL and FEV do as one of its options, i.e., the bisection of an averaged motored curve to produce a reliable determination of the peak position of that curve. In addition, a more sophisticated algorithm unique to this company is offered that calculates the offset correction based on thermodynamic principles. The calculation uses characteristic engine data (bore, stroke, and volumetric compression ratio) along with specific operating parameters (speed, intake temperature, and intake pressure) to derive a TDC result from an averaged, motored curve. This algorithm differentiates diesel engines from gasoline engines. The calculation is aimed at determining the convective exchange coefficient between the combustion gases and cylinder wall. The basis of this calculation is derived from the theory suggested by Hiroyuki Hiroyasu. It utilises the

Eichelberg correlation for diesel engines, and the Woschni correlation for spark ignition engines.

It is proposed that the system calculate the symmetry angle and the TDC offset angle with a 0.01 crank angle degree accuracy (irrespective of acquisition resolution).

General Comment—Motored Curve Method

There are other manufacturers of combustion measurement systems not discussed here, but that may suggest further methods or algorithms for determining TDC. Note that the systems presented here are just a small sample. They do illustrate an important point, though. Even though most suggest a number of alternatives, all manufacturers considered and discussed in the scope of this book offer the bisection-interval method as an option. This leads to an implication that, although this method may not be absolutely perfect, it must provide a good level of accuracy and repeatability without excessive effort, cost, time, or computational power.

It is always interesting to consider and explore alternatives, but it could be concluded from the previous discussion that the bisection-interval method should be considered as first choice when choosing a pressure curve determination method of establishing the correct TDC position, particularly where the target is to start measurements quickly in an efficient way.

7.4.2.3 Capacitive Probe

Measurement of the real dynamic TDC can be established with high levels of accuracy using a specific sensor. This is true because the sensor measures the piston displacement directly and can accommodate the elasticity of the crankshaft. This method can therefore determine the dynamic TDC to an accuracy of 0.1° of crank angle.

Commercially available sensors generally use a capacitive measuring principle: the sensor measures the change in capacitance between the piston and sensor itself. Capacitance changes inversely in proportion to the distance between the piston and sensor. The basic measuring circuit is shown in Figure 7.10.

The sensor probe and the piston are a capacitor. The capacitance depends on the distance between the components. Variations in capacitance are detected by a capacitive bridge circuit. In Figure 7.10, C2 represents the capacitor formed by the sensor probe head and the piston. Square-wave pulses are applied to a capacitive bridge from an oscillator via circuit elements that are dependent on the output voltage. If the capacitance varies, there is a change in the potential divider ratio in the sequence of the capacitors (Cl + C2 or C3 + C4). This change results in a change in the bridge quadrature component of voltage, which in turn causes a change in the output voltage from the differential amplifier and a shift in potential (el and e2). This shift matches the amplitude of the pulses applied via the circuit elements to the new capacitance conditions.

The TDC sensor requires access into the cylinder, which is achieved via an existing bore, for example, via spark plug, glow plug, or injector. Normally, the TDC sensor unit is supplied in a kit, complete with appropriate adaptors as standard equipment. The advantage to this

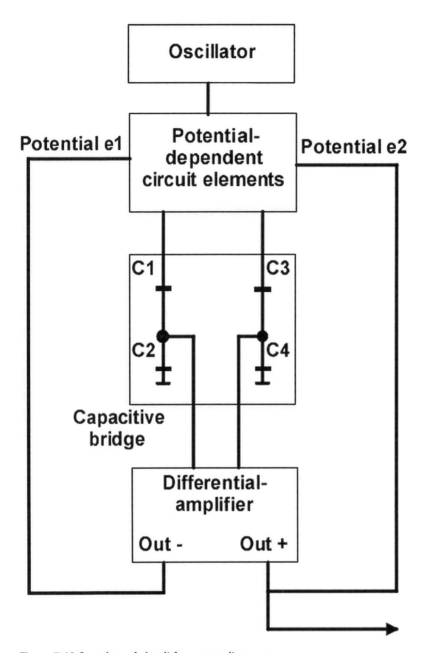

Figure 7.10 Overview of circuit for a capacitve sensor
(Source: AVL.)

technique is that no special machining or preparation of the engine is necessary, because the sensor can be installed and the measurement of TDC effected in a reasonably short time with high accuracy. These considerations make this method an attractive option that offsets the initial cost of purchasing the additional equipment. Typical installations are shown in Figures 7.11 and 7.12.

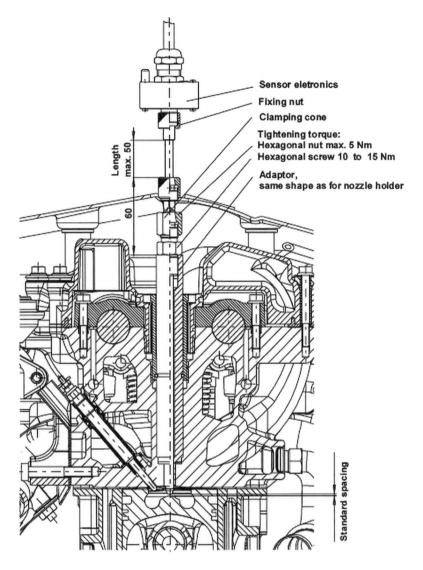

Figure 7.11 TDC sensor mounted in engine via spark plug bore
(Source: AVL.)

This measurement procedure is initiated in a manner similar to that of the motored pressure curve TDC analysis, except of course for using the option for evaluation via TDC probe. The probe produces an analogue voltage signal that is recorded by the combustion measurement system. The engine cylinder is motored via external power or other fired cylinders, and the measurement data are recorded and evaluated by the combustion measurement system. Note that the target cylinder has to be unfired because combustion in the measured cylinder will destroy the sensor probe. In addition, typical engine speeds for evaluation are similar to those used for the motored pressure curve method: Note that

Applications

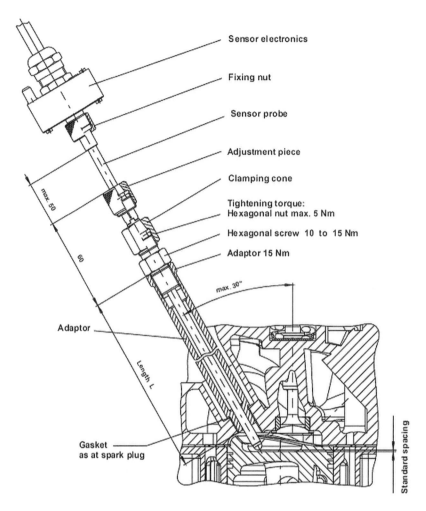

Figure 7.12 TDC sensor mounted via injector installation bore
(Source: AVL.)

above approximately 3000 rpm, the accuracy of the measurement is compromised, and there is risk of damage to the sensor because of the nonrigidity of the engine components under dynamic operating conditions.

The measured curves produced by the sensor are shown in Figure 7.13.

It is interesting to note the reduced amplitude of the curve under compression conditions. That situation is caused by the cylinder pressure acting upon the engine components, thus altering operating clearances. This is a useful phenomenon because it allows discrimination between compression and gas exchange TDC. The correct TDC (i.e., the compression TDC) can then be identified and stored in the parameters of the combustion measurement system. During installation and measurement, the operating clearance of the sensor probe and

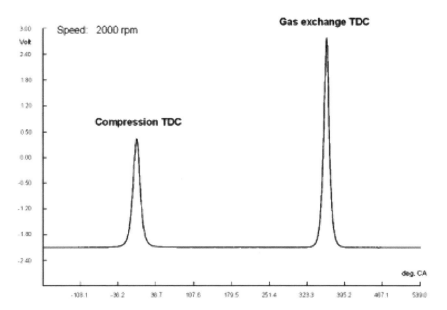

Figure 7.13 Typical measured curves from a capacitive TDC sensor
(Source: AVL.)

piston must be monitored and observed very carefully! There are some general guidelines to be observed during installation of these capacitive type TDC measuring probes:

1. Minimal gap between sensor probe and piston is decisive for a good quality signal, but a certain gap must be present to prevent damage from touching of piston and sensor tip. In general terms, the required gap is between 0.5mm and 1.5mm. The actual value depends very much on the installation position, the angle of the probe, and also the engine type. It is imperative to use the manufacturer's guidelines for setting this gap and to monitor the signal curve while setting up the system and during the measurement.

2. The probe should be as perpendicular to the piston crown as possible. This factor affects the choice of the measuring bore. If possible, the sensor should not be inclined more than 30 degrees from the piston movement axis.

3. If a piston bowl exists, the probe tip should be as close as possible to a central position within the bowl. The edge of the piston bowl must not pass closer than approximately 20mm to the sensor tip. The operator must avoid a setup that, due to the geometry of the piston bowl, causes the smallest distance between sensor probe head and piston to occur before TDC, which would cause the strongest signal to be generated before TDC.

4. Note that certain piston designs could cause problems, particularly those with geometric features above the surface of the piston crown. These are used to generate charge motion, but the transverse motion of these protrusions across the sensor tip can lead to measurement errors.

5. For precision measurement of piston displacement, the measuring position must lie in the same plane of symmetry as the axis of the piston gudgeon pin (also known as a

Applications

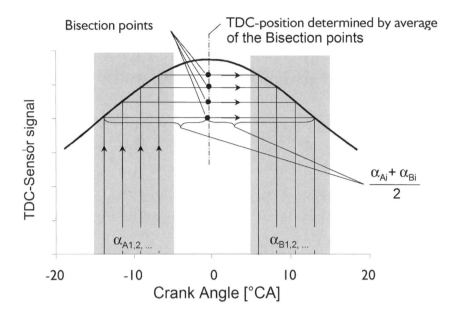

Figure 7.14 Derivation of TDC position from the capacitive sensor signal
(Source: AVL.)

    wrist pin). This alignment prevents any pin offset (implemented by the manufacturer to encourage tilting of the piston around TDC) from adversely affecting the accuracy of the measured result.

6. Although the gas exchange TDC signal is of higher amplitude and could therefore perhaps be considered as better for the establishment of TDC, it is important that the compression TDC be taken, as that is actually more accurate for the purpose than the gas exchange TDC. This is true in part because of the movement of the valves (intake valves are going to open; exhaust valves are going to close) during the gas exchange TDC. This valve movement has a direct influence on the overall system capacitance as the piston moves, resulting in the potential of an incorrect measurement of TDC. During compression TDC, the valves are closed, and therefore, the change of capacitance is linear to the piston movement; this situation promotes a more precise measurement than does the gas exchange TDC.

It is important to remember that the TDC sensor is measuring minimum clearance height, which does not always coincide with 0 degrees TDC. The reason for this are factors such as bearing play and piston pin offset. The analogue voltage produced by the correctly installed sensor gives a smooth, symmetrical curve from which the appropriate TDC position can be established. The analysis generally employed by combustion measurement system manufacturers for deriving TDC from the sensor curve is similar to the interval-bisection method of the motored pressure curve analysis mentioned earlier. That is, the position of the maximum of the curve cannot be used for the derivation of TDC because the piston-to-crank movement ratio is very small around TDC. This method of sensing would be too imprecise. Figure 7.14 portrays the interval-bisection method, a typical approach for establishing TDC from the curve.

7.4.2.4 Microwave

An alternative method for establishing the TDC position via piston movement is to use microwave detection. That technique uses low-power microwaves to establish piston proximity and movement. It is possible to engineer a measurement system, but commercially available systems are available, pioneered by the company Jodon (www.jodon.com).

The equipment comprises a computer-based acquisition system with a number of input channels to allow measurement of crank angle related data, in addition to the real time piston position measurement. The heart of the system is a computer-controlled variable frequency oscillator (VFO), which enables the measurement system to be utilised with a wide range of cylinder displacements. With a VFO range of 8 18GHz, cylinder volumes greater than one litre can be accommodated. The uniqueness of the microwave measurement process is the ability to accurately establish a piston top dead centre (TDC) reference in real time, with the engine running (or cold-motored) at virtually any engine speed. The technique is also applicable to measure timing or phase events on less common reciprocating engine types, for example with rotary or Stirling engines.

The Jodon system eliminates the need to mechanically calibrate the crankshaft rotary motion, or to use damper notch pickups with their inherent inaccuracies. The system is interfaced to the engine using a microwave coaxial cable. Entry into the cylinder head through a diesel probe (IDI or DI version), or through a spark plug adapter (SPA) for SI (spark ignition) engines. Using the SPA, the microwave signal enters the cylinder through a functioning spark plug. A rotational reference input is obtained using a ring gear sensor or engine control system trigger wheel. The system has a number of digital inputs for engine event signals, such as a fuel injector control pulse (SI or diesel engines), a fuel line pressure sensor (diesel engines), or an ignition high-voltage lead sensor (SI). The measurement process can be initiated at virtually any engine speed, and the timing results are presented within a few seconds. With TDC found, and the timing values measured and displayed, the engine speed can be varied and the measurement system will track the speed-related timing events.

If the system were to be used in conjunction with combustion measuring equipment, the angle encoder trigger signal would be measured; the offset between real, dynamic TDC and the trigger signal could then be established with high accuracy. This information could then be entered directly into the combustion measurement system parameter file prior to start of measurement procedures. This task needs to be repeated only for confidence checks, or if the encoder has been removed or disturbed and therefore needs recalibrating for TDC.

7.4.3 Comparison and Discussion of the Methods

Inaccurate determination of TDC, leading to incorrect phasing of crank degree and measured cylinder pressure, is one of the most common errors when technicians execute combustion pressure measurements. As discussed, there are several methods available to establish the correct TDC position of the piston in the cylinder, under dynamic conditions. In addition, there are more rudimentary static methods available. One can conclude, though, that increased cost or effort does result in greater accuracy.

The static method requires minimal expense, but considerable effort. It can be executed in such a way as to lessen the negative effects of working clearances in the engine when measuring, to produce a reasonably accurate result (i.e., by taking both dial gauge readings as the piston approaches TDC). However, this result gives the kinematic TDC, and it does not account for the dynamic effects due to the elasticity of the engine components; in addition, engines with gudgeon pin (or wrist pin) offset have piston motion that is not exactly symmetric about TDC. This fact can cause a small error (0.5 to 1 degree crank angle) in measurement of TDC. The greatest inaccuracy in this method is that, once the TDC position is known, aligning the active edge of the crank angle encoder trigger mark, via manual adjustment, to the TDC position of the engine is subject to human error. Certain optical encoder types allow manual adjustment of the internal disk for this static calibration, but the highest level of accuracy achievable will probably be greater than 0.5 degrees of crank angle.

The motored pressure curve method is a very popular technique for determining TDC. This method is well supported in all combustion measurement systems, and it requires little additional effort and instrumentation cost because it uses the measuring chain that is already installed for combustion measurement. The technique returns a dynamic TDC value that is accurate, apart from heat transfer effects, which are unfortunately unavoidable. This causes the pressure peak to occur earlier than the real TDC, giving rise to the so called *loss angle*—the angle in crank degrees by which the pressure peak leads the actual TDC. This angle has to be estimated or derived, and this estimation leads to the greatest source of error with this technique. The loss angle can be measured, but doing so requires considerable additional effort, and this measurement technique is useful only for the purpose of experimentation. In practical terms, during typical measurement procedures, the loss angle is commonly guessed, or a constant value is always used for certain engine types for repeatability. It should also be noted that the combustion chamber design will have an effect on the loss angle. Certain types of installation in the combustion chamber are not suitable for motored curve measurements. An example would be combustion chambers where the sensor is mounted in the prechamber. The gas dynamics and flow conditions in the prechamber are significantly different from those in the main chamber, and these differences cause unpredictability in the phase shift (i.e., loss angle).

Certain preconditions facilitate a successful motored curve TDC measurement, and these should be observed and applied if possible. The engine should be at normal operating temperature, and, if possible, the measurement should be made by motoring the engine with an external power source (all cylinders unfired—motored via active drive, dynamometer, or starter motor). If the engine is throttled, the throttle should be in the fully open position (if possible) because that throttle position reduces the effects of gas movement and flow between the cylinders that could affect the measured curve. Note that the engine speed during measurement has an effect on the loss angle, assuming this is to be estimated from empirical data supplied by manufacturers of combustion measurement systems.

Measurement of TDC using the sensors described earlier (microwave and capacitive) are highly accurate means of determining a suitable TDC from the minimum clearance height, with no thermodynamic effects. The value returned is generally accurate to within 0.1 degree of crank angle. This is clearly the best method in terms of accuracy. The disadvantage is the additional effort and cost in having to purchase a TDC sensor set and then install it before actually being able to proceed with the combustion measurement task. In addition, the limitations with respect to installation positions may mean that this method is not suitable for application in every engine type.

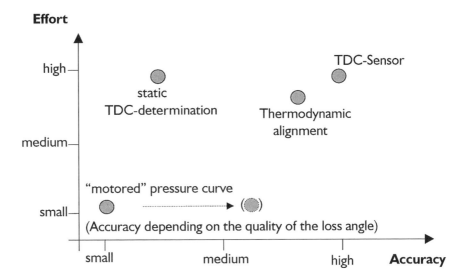

Figure 7.15 Measured TDC accuracy as a function of effort for various TDC determination methods
(Source: AVL.)

One additional method not yet discussed is based on the fact that TDC position can established numerically from the measured combustion pressure curve, compared with simulation data. This involves shifting the curves iteratively to minimise the area difference between them. Heat transfer and combustion-related parameters have to be taken into account, and from this process a calculated motored curve with a simulated TDC position can be established. This method is accurate as long as detailed information regarding the heat transfer properties of that particular engine are available. The errors in establishing the loss angle via this method can be as great as 0.5 degree crank angle if that information is not available.

As mentioned earlier, the accuracy of the resulting TDC value is crucially important to a successful combustion measurement and is one of the greatest sources of error. It is clear that the quality of the final result is a function of the applied effort. The methods we have discussed are shown in Figure 7.15, comparing effort against accuracy.

7.4.4 General Considerations

There are some general considerations to be noted when one is actually executing the TDC measurement, and in particular, dynamic determination methods. In most cases the measurement will be applied to one cylinder, as the reference cylinder. Once TDC is established for that cylinder, the other cylinders will have TDC calculated on the basis of firing order and interval, according to the number of cylinders and the configuration of the engine. In this case, the reference cylinder should be chosen as the one closest to the angle encoder because doing so minimises the effect of the elasticity of the crankshaft affecting the accuracy of the TDC result. This may seem an extreme factor to consider, but we have already discussed the fact that results such as IMEP are very sensitive to any phase error between the measured curve and the crank angle position. In a multicylinder engine it is not uncommon to see crankshaft torsion of at least 0.1 degrees at rated speed and load conditions. The effect can be calculated on the pressure curve. If the torsion of the crankshaft is multiplied by the derivative of pressure with respect to crank angle, the product is the maximum pressure error at that angle. Note that this error has a general trend of increasing with engine speed.

If the encoder and reference cylinder are adjacent, the rigidity of the connection due to the short length and the subsequent errors due to torsion will be insignificant. Note that the errors at each cylinder due to crankshaft torsion will depend on how far the cylinder is from the encoder and how long the crankshaft is (i.e., number of cylinders and cylinder arrangement). The average error is much less than the worst case (i.e., reference cylinder and encoder at opposite ends).

For certain engines, the elasticity of the crankshaft maybe higher than is normally expected. High-performance engines are an example of this. In this case, it could be recommended to carry out a cylinder-specific evaluation of TDC in the individual cylinders prior to the high-pressure measurement. This is particularly appropriate when high-accuracy cylinder pressure data are needed (e.g., for correlation with simulation data). In such a case, measurements of TDC of each cylinder at each engine condition can be made so as to be able to compensate for any effects of engine component behaviour and gas dynamics at operating and measuring conditions. The author has known this to be required in several applications by experience, and an automated measurement procedure, using scripts on the combustion measurement device, is the most effective way to execute this task and subsequently calculate the results.

7.5 Thermodynamic Analysis

7.5.1 Introduction

In the first instance, it is perhaps appropriate to define what we mean by the term *thermodynamic* in the context of this book and this chapter. Engines are, by their nature (to some extent), thermodynamic machines, and they are therefore bound by the fundamental laws of this subject. It is a prerequisite to know this fact for successful handing of a combustion measurement system, in order to generate high-quality data from the in-cylinder measurements.

We have previously discussed the fact that the combustion measurement system is typically used for measuring in-cylinder pressure during engine operation. The system measures crank degrees, and it correctly aligns this measurement with the pressure curve so the relationship between them is accurate to a high degree. From the crank displacement, instantaneous in-cylinder volume can be calculated with a simple, well-established algorithm. From these readings we have the pressure and volume information from which further calculations, of varying degrees of complexity, can be derived, and it is a common requirement that these calculations be executed during the measurement.

These calculations derive results from the pressure curve that are of interest to the engine engineer. In addition, the extraction of significant results in real time is a method of data reduction that helps us gain an understanding of the in-cylinder process, in a concise way, as opposed to handling large, cyclic curve datasets. We have already discussed the fact that these results fall into two main categories—indirect and direct results. Direct results can be derived directly from the measured curve, e.g., the position of maximum pressure. Indirect results need further information relating to pressure and volume for us to derive the output, e.g., the IMEP.

Thermodynamic results fall into the latter category (indirect results). They are more sophisticated calculations to derive energy released by the cylinder charge. IMEP is a calculation that represents the work done in each cycle. Thermodynamic calculations commonly return the instantaneous energy release that allows the engineer to understand the progress and rate of thermal energy conversion in the combustion chamber. These data are normally expressed as crank angle based curves (derivative and integral curves). These curves are used for the extraction of specific results that are essential knowledge when we are trying to understand the behaviour of the engine and combustion process. An example is the point of 50% energy conversion because that has great significance for a gasoline engine, or the start of combustion, which has great significance for Diesel engines.

These calculations of energy released versus crank angle fall into two main categories. These are commonly known as *burn rate* and *heat release*. The former is normally applied to and associated with gasoline engines, and it produces a normalised energy release curve with a scale of from 0 to 100%. The so-called *burn angles* or *energy conversion points* are extracted from this curve (also known as *mass burned fractions*—MBF). These data are of great interest to the engine engineer because they are typically used to measure the start of combustion (5-10%) and the end of combustion (90-100%). In addition the 50% mass burned fraction is an important calibration strategy target because of its direct correlation with MBT (minimum spark advance for best torque).

The latter—heat release—is commonly associated with diesel or compression ignition engines. The energy release curve has a scale of absolute units of energy, and it can be compared with the chemical energy introduced into the cylinder via the injection quantity. The results can then be studied in detail to understand the energy conversion and associated efficiencies of the processes taking place in the combustion chamber.

The derivation of energy release from the cylinder high pressure curve is well established theory (discussed next) even though the principles were developed many years ago, they are well proven and accepted, and have stood the test of time.

The algorithms that can be applied vary from relatively simple to very sophisticated. Each additional level of sophistication requires more parameters and therefore requires more processing time. This compromise should be considered carefully because simplified algorithms can give excellent, repeatable results because of minimal assumed parameters. The additional complication of a detailed analysis may not be possible to execute during the measurement, and it may not add any value to the measurement task. In order to judge this compromise, the user must be familiar with the scope of available methods and compromises. The following explanation is an attempt to provide the basic knowledge that can by supplemented by further research so that users can make the correct choice for each of their tasks.

7.5.2 Basic Principles and Early Work

The most famous and well-accepted work on the topic of instantaneous energy release derived from the pressure curve was carried out by Messrs Rassweiler and Withrow in the mid 1930s. They produced a paper of experimental work carried out using high-speed photography, showing successive positions of the combustion flame of a single engine cycle. This was compared with cylinder pressure data of that cycle with respect to time and crank angle.

In summary, their work proposed a method for discriminating between the pressure change in the cylinder due to combustion, and the pressure change due to volume change. Once the pressure change due to the combustion event was separated from that caused by volume change, it was found that, after summing and normalising this value, it had direct correlation to the percentage of charge burned (by weight) at the corresponding crank or time positions.

Their setup and the description of the experimental process are written up in detail in the appropriate technical paper, and those details go beyond the scope of this book. However, a discussion follows here of the explanation of the basic theory of the correlation of pressure and mass burned. This discussion will give the reader sufficient detail to understand the basic theory, which can then be supported by further research.

Rasswieler and Withrow compared data from cylinder pressure measurements with the high-speed motion pictures taken from the engine combustion chamber. Due to the very small change in cylinder volume around TDC, they proposed that one could assume that the combustion event takes place at constant combustion chamber volume. Based on that assumption, they proposed that the engine combustion process could be compared to the process that occurs in a bomb calorimeter (also at constant volume). Other experimentation highlighted that, in the bomb calorimeter, the fraction of the mass charge burned at any instant during combustion has a direct correlation to the fractional pressure at the same point. Consequently, if we assume that the engine combustion process takes place at constant volume, it is reasonable to expect that the observed cylinder pressure curve (due to combustion only) would be the same shape as the energy release curve, when plotted with respectively correct scales.

In order to derive this pressure curve based on combustion only, the pressure change due to volume change must be removed from the cylinder pressure curve. The total pressure change due to combustion and volume change is measured via the cylinder pressure sensor. The pressure due to volume alone can be derived by knowing the change in volume and the polytropic exponent that describes the process during the volume change (often also describing the state of the gas). This exponent can be derived during experimentation, but generally a known or accepted figure can be used (Rassweiler and Withrow used 1.3). With this information, the pressure change between 2 points can be identified separately as that caused by volume change and that caused by combustion only.

Once the pressure rise due to combustion is separated, we must consider what the pressure rise would have been had it occurred at constant volume. Let's go back to the bomb calorimeter. The pressure increase in a constant-volume bomb calorimeter, by creating a certain amount of heat, is inversely proportional to the total volume. This means that the increase in pressure caused by combustion only can be normalised to a constant, volume-only pressure increase by multiplying it by the ratio of the actual cylinder volume, divided by the constant reference volume. This reference volume is normally the volume at the start of combustion or at TDC (minimum cylinder volume). The process is carried out in incremental steps through the pressure versus volume crank angle data set to produce a pressure curve that has been normalised with respect to pressure changes from volume changes, and that is referenced to a fixed volume. This curve now approximates very closely the energy conversion rate in the cylinder, and it can be used as a measure of the rate of combustion. The following point should be noted, though, with respect to this calculation:

1. The smaller the incremental steps for the calculation the better the approximation, although such incrementalisation increases processing time.
2. The effects of heat loss are mostly neglected if a constant polytropic exponent is used throughout the cycle for calculation.
3. Any leakages in the cylinder will affect the quality of the resultant data.

Although this method has some shortfalls, it is generally accurate, particularly in relative terms; in addition, the method does not require extensive calculation effort. For this reason, the theory is well established in industry, irrespective of the fact that capabilities of modern data acquisition systems now far exceed those available to Rassweiler and Withrow. This theory is still used as the underpinning methodology to much more complex calculations that are available for use today. In addition, the simplicity of this calculation means that it forms the basis for the real-time calculations for energy release used on many modern combustion measurement systems because it is so computationally efficient, as well as giving repeatable results.

7.5.3 Methods for Real-time Analysis

One of the advantages of using a combustion measurement system is that the heat release curves and results can generally be available in real time. In the past, data had to be measured, stored, and then analysed before any of the energy-conversion rate curves or results could be derived. The reason for this is that the computational power of systems in the past limited them to data acquisition only. That is, there was enough processing power only to acquire the data and digitise it in real time. With developments in processor power and system architecture, the possibility to hard-code heat release calculations and execute them in a deterministic way during measurement became reality. These systems have evolved on an ongoing basis, and now heat release calculation is a standard function for any combustion measurement system.

A basic algorithm for fast heat release is proposed and implemented by AVL and is stated to be based on the first law of thermodynamics:

This calculation is based on a simplified process, and it returns the rate of heat release, which represents the energy effectively delivered to the gas from the cylinder pressure. The surface losses are completely neglected, and, as a result, the displayed net heat release is accordingly

Applications

$$Q_i = \frac{K}{\kappa - 1}\left[\kappa \cdot p_i \cdot (V_{i+n} - V_{i-n}) + V_i \cdot (p_{i+n} - p_{i-n})\right]$$

n ... interval (1 deg. CA)

κ ... polytropic coefficient

p ... cylinder pressure

V ... volume

K ... constant

Figure 7.16 Fast heat release according to the first law of thermodynamics law
(Source: AVL.)

lower than the actual gross energy released (by approximately 20%). Note that the calculation does not discriminate between the exponent that describes the compression and expansion process, and the exponent that describes the state of the gas. They are assumed to be the same, and they are given by a parameter setting that the user selects during parameterisation of the system. This simplifies the calculation so that it can be executed more quickly, but, obviously accuracy is less in absolute terms. For most studies, though, repeatability is the most important factor, and a simple algorithm such as this will give good performance in that respect.

A development of this basic algorithm became available as processing power in combustion measurement systems increased. This increase allowed more parameters to be set according to the application in order to increase accuracy and give a higher degree of freedom to the user to tune the hard-coded algorithm to suit the specific engine or test target. A typical example is shown above. This is AVL's *Thermodynamics 2* calculation. It is again a simplified calculation that returns the energy delivered to the gas, ignoring surface losses. The result is therefore net heat release (as opposed to gross heat release). The main difference is that, in this calculation, a variable polytropic coefficient is taken into account during the cycle calculation. This is maintained in relation to the temperature by an approximation formula. In order to determine the bulk gas temperature, though, the gas mass must derived or known. The gas mass is taken into account in the algorithm, and the mass is calculated from intake manifold pressure, intake manifold temperature, and volumetric efficiency.

The heat release algorithm is is displayed in the following equations:

V_H swept volume
p_s intake manifold pressure
R univ. gas constant = 287.12 kJ/kg · K
T_s intake manifold temperature

$$Q_i = \frac{C}{\kappa_i - 1} V_{i+n} \cdot \left[p_{i+n} - p_{i-n} \left(\frac{V_{i-n}}{V_{i+n}} \right)^{\kappa_i} \right] \cdot (X_i + 1)$$

κ_i polytropic coefficient
Cv_i spec. heat at constant volume
T_i temperature
n   interval (1 degree)
C   constant

$$\kappa_i = \frac{0{,}2888}{Cv_i} + 1$$

$$Cv_i = 0{,}7 + T_i \cdot (0{,}155 + A_i) \cdot 10^{-3}$$

$A_1 = 0.1$ for gasoline engines
$A_1 = 2X_i$ for diesel engines

$$X_i = \frac{\sum Q_i \cdot 28}{p_s}$$

$$T_i = \frac{p_i \cdot V_i}{m \cdot R}$$

The gas mass calculation is given here:

$$m = l \cdot m_{th} = l \cdot V_H \cdot p = l \cdot V_H \cdot \frac{p_s}{R \cdot T_s}$$

l	volumetric efficiency
m_{th}	theoretical air mass
ρ	gas density

More sophisticated variations of the heat release calculation are available that provide a more accurate calculation of energy release curve and associated results. Care should be taken, though, because these calculations have many parameters that must be known and that must be set correctly. If realistic values are not available for these calculation parameters, then there is no benefit in using a more sophisticated calculation that involves the use of valuable processing resources. As the calculations become more complex, with more parameters to set, they are of course more demanding to execute in real time. When one is using these more sophisticated versions of heat release calculations, care must be taken to understand clearly where the calculation will be executed and how deterministic this process is. The fast, real-time processors typically used in combustion measurement systems for executing real-time calculations have limited memory on board, and they are not capable of manipulating complex algorithms with many parameters. It may be that a sophisticated calculation for heat release is actually running on the computer user-interface hardware, in which case the calculation of the result may not be returned in real time. For steady-state testing, this is acceptable, whereas for transient mode testing, particularly for the development of engine control loops under these conditions, the calculation time may be too long.

7.5.4 Further Discussion—Offline Analysis

A more detailed study of the pressure curve information, and of the resulting heat curve calculations, will often be carried out offline as a postprocessing exercise. This gives the engineer the possibility to use more complex, alternative algorithms for the calculation of heat release.

There are many published alternatives to the more basic algorithms executed in real time. In addition, the crank angle data sets can be averaged to minimise the effects of cyclic variations in order to try to achieve a more objective calculation for the energy release curve from which the conversion points can be extracted.

Most of the more sophisticated calculations are more complex versions of the basic Rassweiler and Withrow algorithm that have factors or parameters to overcome the shortfall of this simple algorithm that was proposed in a time when measurement equipment of the sophistication we have at the time of this writing was not available. In particular, these algorithms take into account the errors introduced by imprecise definition of the polytropic index, and they can account for heat loss through the cylinder walls and head.

7.5.4.1 Wiebe Function

When considering mass burn fraction curves derived from simple heat release calculations, we note that these often have a similar characteristic shape and profile that is often modelled or expressed in terms of a calculation. Often this calculated curve is used as an input into further simulations. The mass burn curve is often expressed by a function is known as the *Wiebe function*. It gives the cumulative mass fraction burned as a function of crank angle, as explained here:

$$m_b = 1 - \exp\left[-a\left(\frac{\theta - \theta_s}{\Delta\theta}\right)^{m+1}\right]$$

θ is the crank angle

θ_s is the crank angle at the start of combustion

$\Delta\theta$ is the duration of combustion

'a' and 'm' are user defined parameters

The parameters can be adjusted to give the required shape and form of this curve. Typical values for parameters a and m are 5 and 2 respectively. A comparison of methods is shown in Figure 7.17.

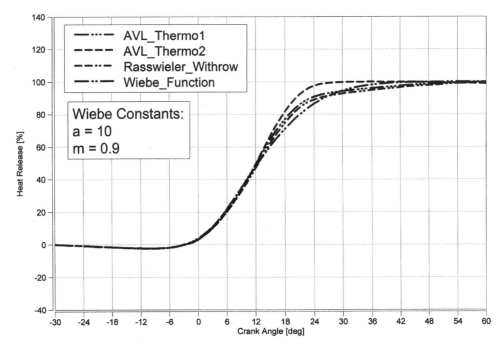

Figure 7.17 A comparison of heat release calculations

(Source: AVL.)

7.5.7 General Comments

As mentioned previously, it is important to really define the objective of the data produced by any algorithm. Highly accurate calculation of heat release is generally required for laboratory or experimental tests. For most routine engine development tasks, the most important factor is to have a reliable and repeatable calculation that can be executed quickly to produce the necessary data.

General errors relating to the measurement technique can cause a number of effects, specifically with respect to the derivation of energy release from the pressure curve. Some have greater effect than others on the calculations, and these should be considered when setting up and parameterising the combustion measurement system, or when analysing the data. Here are some important, error-producing effects to consider:

Zero-level correction errors—The piezoelectric measurement chain, being a dynamic measurement technique, needs correction of the measured curve to achieve an absolute pressure curve. The techniques for this have already been discussed in an earlier section, but it should be noted that correct establishment of the absolute pressure for the measured curve is essential for a correct derivation of the energy release curve. This is true because small errors in the pressure referencing create large errors in the calculation of the polytropic index that defines the compression and expansion process. This is particularly significant at engine light-load or part-load operating conditions

Signal noise—Noisy signals are clearly undesirable. The effect of noise spikes on the pressure curve, particularly in the initial or latter stages of energy conversion, is to produce even noisier heat release curves, from which the results cannot be derived effectively or repeatably. This is a particular issue at low load–slow burn conditions. It is common that manufacturers of combustion measurement equipment include selectable filter algorithms as parameters in their heat release calculations.

TDC errors—The effect of crank angle phasing errors is much less on heat release calculations when compared to IMEP. However, incorrect pressure versus volume relationship does have an effect on the slope of the integrated heat curve, causing small errors that are generally insensitive to engine load conditions. In addition, incorrect TDC has a small effect on calculated indexes for the compression and expansion processes.

Engine geometry—Errors in this area affect the quality of the complete measurement procedure, not of just the heat release calculations. However, these calculations show particular sensitivity to correct determination of the compression ratio. An incorrect definition causes errors in particular around TDC, where the largest volume ratio errors exist.

In summary, the quality of the procedures around setting up and executing the combustion measurement will have a significant effect on the quality of the heat release data (as well as on other data from the measurement). The points that are particularly important for heat release calculations are these:

> Correct and accurate zero-level correction
>
> The effects of abnormal combustion (misfires, slow burns)

Accurate definition of start and end of combustion

Good quality pressure data, particularly at the beginning and end of combustion

7.6 Low Pressure Measurement and Gas Exchange Analysis

7.6.1 Introduction

Often, a detailed understanding of the low pressure-gas exchange process is needed to be able to fully understand and optimise the complete engine cycle. The purpose of the gas exchange part of the cycle is to remove the burned gases from combustion after the power stroke, and to replace them with a fresh charge ready for the next cycle. The most important element of the gas exchange process is to induce the maximum charge mass into the cylinder (with respect to the actual operating conditions) and to retain it for the combustion process as efficiently as possible.

There are many factors that can assist or impede this process, apart from those relating to the combustion chamber itself (piston, valves, combustion chamber design, etc.). The external subsystems around the engine that fulfil the task of managing gas flows into and out of the engine are of particular significance with respect to improving the efficiency of the in-cylinder processes. The behaviour of these components and their effect with respect to the cylinder gas exchange is of great interest to the engine engineer. In addition, many engines are equipped with devices to increase the airflow through the engine (turbochargers, etc.) to improve power density, and with devices to aftertreat emissions and to control combustion. All these systems have an effect on the efficiency of the gas flows into and out of the cylinder.

Low pressure measurement (Figure 7.18) is also often used in the detailed study of the gas exchange processes, in addition to evaluation of engine friction losses. The MEP (mean

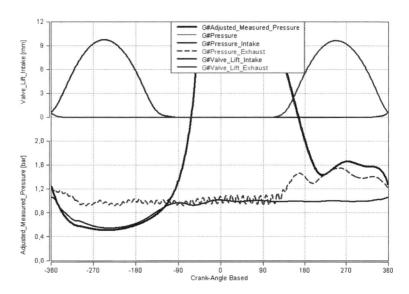

Figure 7.18 Low pressure curves over one cycle

(Source: AVL.)

effective pressure) associated with the gas exchange process (often known as PMEP—pumping mean effective pressure) can be compared to the BMEP (brake mean effective pressure), and this comparison permits the establishment of losses that can be apportioned directly to friction and auxiliary drives.

The low pressure measurement has become of particular interest as a specialised measurement for the combustion measurement system. It provides invaluable information that can be utilised directly for improving the overall efficiency of the engine.

7.6.2 Measurement Task and Goal

The basic requirement of the low pressure measurement is to gather accurate pressure data from the inlet and exhaust manifold, from as close as possible to the engine valves. In addition to the low pressure measurement, the high pressure part of the cycle is also measured to complete the information needed for the cycle analysis. Typical measured curves from a low pressure analysis are shown in Figure 7.19.

During optimisation of the gas exchange process for internal combustion engines, the tasks commonly comprise pressure measurement, calculation of heat release, and close examination of the gas exchange process. This work is appended with further calculations and verified via engine simulation code.

These techniques, when used in combination, can significantly reduce the time needed to develop an optimized gas exchange process in an engine. Generally a complete pressure analysis, including a full analysis of the combustion and gas exchange processes is necessary

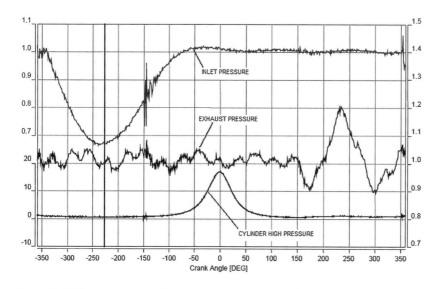

Figure 7.19 Measured curves from a low pressure analysis
(Source: AVL.)

for calculation and derivation of the residual gas masses and for an accurate rate of heat release.

In addition, the simulation is used to verify the quality of the pressure measurements. The detailed analysis can be used to correct errors and to calibrate the output of the measurement chain. This information is essential for the development of the camshaft geometry and valve lift profiles.

A typical method of analysing the working process of combustion engines is by means of relatively simple energy balances according to the first law of thermodynamics. This is used for the basic assessment of the base engine design.

For the accuracy of this process, establishing the initial conditions of the inlet charge with respect to mass and composition is of fundamental importance. This is normally determined by a gas exchange calculation carried out using a flow equation and the intake and exhaust pressure curves, which are measured via the low pressure measurement and can be verified via simulation.

In order to reduce emissions of HC (hydrocarbons) and NOx (oxides of nitrogen), it is common practice to use internal exhaust gas recirculation. Such recirculation is particularly effective at part load, and it does not require any external sensors, pipe work, or equipment on the engine. The exhaust-gas recirculation is controlled by the gas exchange process, and this has the effect of increasing the residual exhaust gas content in the cylinder, reducing the combustion temperatures, and thus reducing NOx emissions. In addition, the higher average HC concentration of the residual exhaust gas, compared with the exhaust gas, reduces the HC emissions.

The exhaust gas content of the in-cylinder charge also influences the thermodynamic efficiency of the complete engine cycle. This is so because the higher specific heat capacity of the residual exhaust gas, during compression, reduces the isentropic exponents of the charge and thus the compression work. In the expansion phase, with a consistent air-fuel ratio, the reduced fuel mass results in lower process temperatures, and in a corresponding rise in the isentropic exponents. These two factors have a cumulative effect, and they generally lead to an overall increase in efficiency, unless the residual exhaust gas content is so high that there is a deficiency of air.

A common method of controlling exhaust gas recirculation in modern engine designs is via variable valve timing, generally applied to spark ignition engines. This is an efficient way of controlling the internal EGR (exhaust gas recirculation) rate. Using this technique, a manufacturer can claim an overall increase in internal efficiency of up to 10%.

Internal EGR is impossible to measure directly. The value must be derived from the result of simulation, but the simulation must be proven accurate via measurement. Optimisation of the gas exchange process to give an understanding of internal EGR rate, plus trapped masses and efficiencies, can be achieved via simulation. The prerequisite for this is that the simulation model needs an accurate determination of the residual exhaust gas content at the moment when the intake valve closes as an important initial condition for the working process calculation. Thus, the low pressure measurement must provide accurate data with respect to the low pressure part of the cycle, and particularly with respect to the absolute pressure level.

7.6.3 Typical Measurement Setup

The measurement set up for a typical low pressure analysis measurement involves installation of transducers in the inlet and exhaust tract. These must be installed as close as possible to the valves, in addition to having the usual combustion chamber-mounted transducer. Often only one cylinder will be instrumented, even in a multicylinder application, although there is no technical reason why multiple cylinders cannot be observed. Figure 7.20 shows the general arrangement for a cylinder.

Low pressure measurements place extremely high demands on the sensors. The operating conditions are in some respects more severe than those encountered in the cylinder. Vibration levels are high because the sensors are mounted into components with relatively low mass when compared to the engine itself. This means that the possibility of vibration frequencies being induced onto the measured curve is high. In addition, the range of the measured curve does not have a high amplitude (when compared to a high pressure measurement). The impact of this is that the signal is more sensitive to noise because the signal-noise ratio will be less favourable in the application. The chosen sensor must therefore have high sensitivity, which in turn increases its physical size, and thus makes it more difficult to accommodate near the ideal measuring point at the back of the engine valve.

The sensors for the low pressure measurements will be subjected to high temperatures, particularly on the exhaust side, but also on the inlet side, particularly for pressure-charged engines with external EGR. The gas temperatures of more than 700° C in the exhaust system

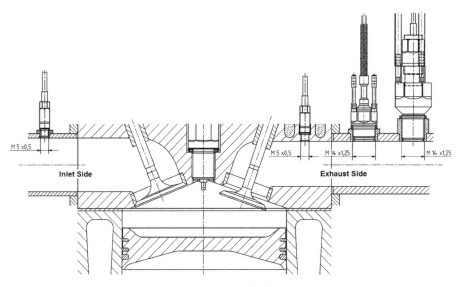

Figure 7.20 General arrangement of transducers for low-pressure measurement
(Source: Kistler.)

mean that for currently available transducer types, it is essential that they be mounted into a cooling adapter or, even more favourable, into a cooled, switching adaptor that prevents the sensor from being exposed to the hot gas any longer than is actually necessary for the measurement. The disadvantage of these adaptors is that this means that they have to be mounted at a considerable distance from the exhaust valve due to their physical size.

Generally for low pressure measurements, absolute measurement technology is most favourable. It is essential that the low pressure measured curve, on both inlet and exhaust sides, be available as an absolute pressure measurement. This data from the inlet side is often used in the simulation as the reference for correcting the cylinder pressure curve. It can be stated then that for this application, piezoresistive transducers are favoured, although they are not the only choice. It is possible to use piezoelectric sensors of course, but these need an additional absolute measuring sensor, used in combination, to establish the absolute pressure value.

It is particularly important that the transducers used for low pressure analysis are as thermally stable as possible because sensitivity shift due to temperature change can have a negative affect on the residual gas content calculation. In addition, the gas exchange analysis reacts particularly sensitively to a shift in the exhaust pressure level. A shift in the measured pressure level also affects the calculated mass flows in the intake and exhaust systems. Similar effects can be apparent on the calculated cylinder mass, which reacts significantly more sensitively to a shift in the pressure level in the exhaust system than to a corresponding shift in the intake system. It is clear, then, that analogous to other calculations carried out during a typical combustion measurement, small errors in the acquired data produce large errors in the calculated or derived results.

7.6.4 Measurement and analysis

Detailed analysis of the gas exchange process, in combination with simulated data, is becoming a powerful tool in the development of modern engines. In particular this measurement can be used to give a detailed analysis of the cylinder charging process and of the efficiencies associated with that process. Modern engine technology dictates that certain values must be available for correct calibration of control units. These values are cannot always be achieved via direct measurement. Rather, they are values that are derived from a detailed numerical analysis of the engine cycle. That fact is the main driving force behind the need for low pressure measurement as a standard measurement technique at the engine test bed during the routine development process, as opposed to being generally found in fundamental combustion research laboratory environments.

The company AVL has developed an interesting approach to bringing simulation and measurement closer together. The concept involves the use of a simplified, single dimensional engine combustion simulation model. This is used in conjunction with the measured low and high pressure curves to return the required calculated values. Of particular value are the residual gas content and the trapped masses and associated efficiencies, in addition to a detailed energy balance calculation.

The simulation model used is a simplified, one-dimension, gas dynamic cyclic simulation model of the cylinder and ports that utilises a combustion model based on rate of heat

Applications

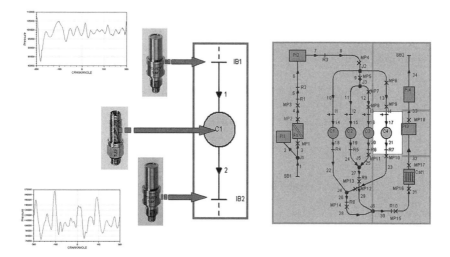

Figure 7.21 General principle of reduced simulation model for gas exchange and combustion analysis online
(Source: AVL.)

release from combustion analysis. The general principle is shown in Figure 7.21. The simulation code can be dramatically simplified due to the low pressure measurement data. Assuming that the gas pressure and temperatures are measured, the flow dynamics and masses can be calculated, once the valve lifts and opening properties are known, in addition to the flow coefficients.

The simulation is executed immediately after the steady-state measured data are recorded and stored. Calculation time is typically 30 seconds for a measurement of 200 averaged cycles. This means that resulting, calculated values are available to the engineer within a minute or so of the actual measurement being taken. The advantage is that any discrepancies, errors, or data validation via additional tests can be dealt with immediately if needed. The general process of the procedure is shown in Figure 7.22.

Typical curves and results returned are shown in Figure 7.23.

7.6.5 Summary

The most important aspects to consider for the low pressure–gas exchange measurement procedure is in relation to the low pressure transducers and their installation, as this is the only part that is specific for the measurement. The rest of the equipment is the standard combustion measurement chain, and thus the general guidelines already discussed in previous chapters apply here. The main considerations for the low pressure transducers and their installation are these:

If possible, absolute measuring transducers should be chosen in preference over low pressure measurements.

Low Pressure Measurement and Gas Exchange Analysis | Chapter 7

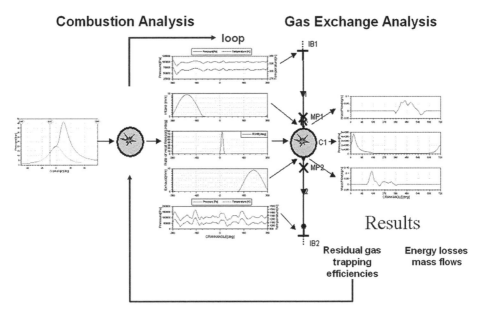

Figure 7.22 Measurement and Simulation process for low pressure analysis and detailed cycle calculations
(Source: AVL.)

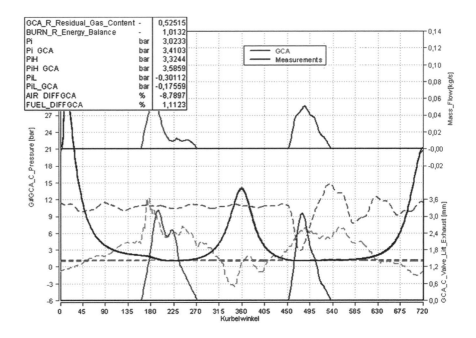

Figure 7.23 Example of results from combined measurement/simulation procedure
(Source: AVL.)

215

The transducers should be placed as close as physically possible to the engine valves; this is particularly important if the calculation model does not account for gas dynamics.

The transducers will generally require adaptor mounting to protect them during operation, which limits the possibilities for fulfilling the preceding requirement.

The transducers should have stable properties with respect to the temperature changes (in respect to sensitivity and zero-level shift) and should be accurate to within ± 10 millibar for accurate zero-level correction of the high pressure curve for detailed cycle analysis.

The transducers should have low sensitivity to vibration.

Once the measurement setup is complete, additional information is needed for parameterisation of the calculation model for gas exchange, namely for the parameters that describe the operational characteristics of the engine valves—opening curves and flow coefficients.

Chapter 8
Abnormal Combustion Measurement and Evaluation

8.1 Combustion Knock and Abnormal Combustion (SI Engines)

8.1.1 Introduction

Knock is a common term used to describe an abnormal combustion phenomenon in gasoline engines that generally produces a characteristic noise under certain engine operating conditions. There are a number of terms associated with this phenomena that confusingly may describe different phenomena but that all lead to the same result, i.e., the knocking or pinging sound. Even more confusing, certain of these terms effectively describe the same things:

- Pre-ignition
- Post-ignition
- Self-ignition
- Surface ignition
- Detonation
- Pinging

Additional confusion comes because experts do not always agree on the terms used to describe details of the phenomena. In an attempt to provide some clarity within the scope of this book, a description of abnormal combustion for gasoline engines will be made; terms described will remain consistent throughout the text.

8.1.2 What Are Abnormal Combustion and Knock?

First, it is important to define normal combustion in a spark ignited engine: combustion of the fuel/air mixture, initiated at the correct time, and solely via the ignition system, that produces a controlled, predictable burning and energy release from the in-cylinder charge. Regular combustion is controlled via the spark advance, and via charge state (turbulence).

Abnormal combustion can refer to a variety of situations in which normal combustion does not occur. This includes modes wherein the flame fails to consume all of the charge—known as partial burns, or misfires—as well as conditions such that the flame front is initiated before or after the timed spark ignition by other means (i.e., pre- or post-ignition) or wherein some or all of the charge is combusts spontaneously. This latter effect produces uncontrolled burning and high, instantaneous energy release rates from the charge. The consequence of this is very high localized pressure rises that produce the damaging, high-amplitude pressure waves that impart the acoustic resonance known generically as knock.

There are a number of situations where knock can occur. Some can be controlled, but others are to be avoided, for they can cause significant engine damage in a very short time. Figure 8.1 maps the causes of irregular combustion.

Spark knock is a common phenomena and causes a compromise in the fuel efficiency of the engine. When spark knock occurs the end-gas region of the combustion chamber, is compressed by the advancing flame front to pressures and temperatures sufficiently

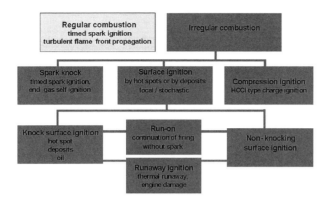

Figure 8.1 Normal and abnormal combustion in a gasoline engine.
(Source: Heywood.)

high to initiate spontaneous combustion of these gases. This self-ignition event occurs spontaneously and provides a rapid, uncontrolled release of heat energy. This creates very high localised pressure gradients that excite resonances in the combustion chamber, thus producing the damaging subsonic pressure waves that create the characteristic noise and physical damage.

This type of self-ignition event can occur repeatably during a normal combustion event, particularly at the limit of ignition advance. It can be controlled via adjustment of the ignition firing angle, and it is often a factor that is associated with fuel quality and fundamental combustion chamber design. Spark knock is important, for it is a constraint on engine efficiency by effectively limiting the maximum compression ratio with which the engine can run when using a specific fuel.

What we can say is that irregular combustion caused by spark knock is a function of the state of the end gas with respect to its temperature and pressure. Another cause of self-ignition and knock is a pre-ignition event, so called because it causes ignition and combustion to occur before the main spark timed event. This cause of knock has become of great interest recently because of modern engines' high specific power output and consequent high thermal load. Such engines are susceptible to self-ignition of the charge, prior to the ignition event, from glowing particles or deposits that can appear in the combustion chamber. These deposits often form on the back of the inlet valves, then break free and are induced into the combustion chamber, where they may pass through, or may remain during some operational cycles, then being heated to a temperature that is sufficient to ignite a fresh charge when compressed. This type of ignition mode causes the rapid, uncontrolled energy release with the associated acoustic noise and physical damage. These events are generally short and of high intensity and can cause terminal damage over a few engine cycles; they are often known as super knock or mega knock events, the effect on the measured cylinder pressure curve is shown in Figure 8.2. Engine developers try to fully understand and avoid this phenomena during the design and testing of new engine types.

In summary then, knock is a damaging phenomena that is caused by uncontrolled combustion; this can be initiated in a managed way (spark knock) or can occur in an uncontrolled way (glowing self ignition).

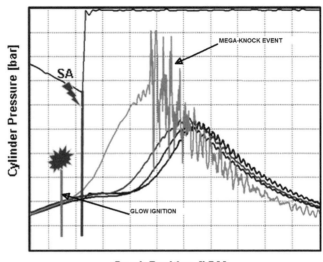

Figure 8.2 Pressure curve showing typical mega knock event.
(Source: AVL.)

During normal combustion, the flame front progresses outwards radially from an initial kernel at the spark plug at a velocity of up to 50 metres per second. This process is generally controlled and predictable, and the pressure in the cylinder is approximately the same at any point (i.e. an average cylinder pressure). During a knock combustion event, the natural frequencies of the combustion chamber are excited; this results in pressure waves that propagate around the combustion chamber at much higher speeds, up to 1,000 metres per second depending on the state of the combustion chamber gases.

The resonance and subsequent oscillations, which can be observed clearly on the cylinder pressure curve, provide a reliable indication of the occurrence and severity of the knock event. Thus, the combustion measurement device may easily be applied for measuring and monitoring knock events in spark ignition combustion engines.

8.1.3 Why Is Knock Important?

Knock is an important phenomena that needs close observation during the engine development process. In the early stages, the engineer developing the fundamental design of the combustion chamber must ensure that the design shows some resilience to knocking, particularly when operating using a wide range of fuels, all of which may be utilized in an engine designed for a global market. In order to achieve the highest thermal efficiency for the engine cycle, the combustion system must operate with the highest possible compression ratio. The limiting factor for this is the engine's knock characteristic, or its tendency to knock

Later on, in the development phase, the calibration engineer must be aware of knock while tuning the engine control system, often in the vehicle. To achieve maximum efficiency, an engine must run with ignition timing angles that are close to the spark knock borderline. And because knock is unpleasant for drivers, even at low levels, knock avoidance is even more desirable.

The chemistry of the fuel used in the engine has a significant effect; the fuel's resistance to knock events is described by its RON (Research Octane Number). Because the characteristics of the fuel can be improved with the addition of additives, establishment of knock via measurement is particularly important for fuel development work, as well as to fuel scientists.

There are certain engine operating conditions under which knock is most likely: generally high load conditions, particularly at low speed. Combustion chamber deposits also have an effect by effectively causing glowing hot spots; in addition, deposit layers in the combustion chamber reduce the heat transfer capability of the surfaces, thus causing higher combustion chamber temperatures. Furthermore, the volume of these deposits effectively reduces the compression space, thus increasing the compression ratio and thereby the tendency to knock when the engine in running using a given fuel quality and at a certain ignition timing.

Elevated temperatures also affect the combustion chamber—overall engine temperature as well as ambient temperatures. Of course, correct spark timing for the engine operating condition is also an influence.

The occurrence of knock is not just annoying because of the noise associated with it; even more important, it is terribly destructive to the engine—and significant costs are associated with getting rid of it. Improved fuel quality can help, but the cost of such an approach is generally prohibitive. Changes in engine setup can reduce the tendency to knock, as can reduced compression ratio, changes in the spark timing angle, and a richer mixture, but such solutions degrade the overall efficiency of the engine, reducing power while increasing fuel consumption.

The main objective, then, is to design a combustion system that is has low tendency for knock and that can thus run at the highest possible compression ratio with optimal ignition timing, achieving maximum performance and efficiency.

8.1.4 Knock Measurement and Analysis

8.1.4.1 Knock Measurement Techniques

The most important aspect of knock with respect to the combustion measurement is how can it be measured accurately and repeatably. There are many general approaches to knock detection—one of the most surprisingly sensitive is the human ear. Hearing is an important factor, for the user of the engine will, in general, only be aware of knock because of the noise that it generates.

For production engine control systems, structure-bourne knock sensors are often used; these can be applied as a single sensor per engine, or multiple sensors in groups. The signal is interpreted via a specific window relating to crank position that allows it to be associated with a firing cylinder. The reason for this is that the knock event only occurs in a specific crank degree range during the cycle. These sensors are tuned to the knock frequencies of the engine and resonate when knock occurs, producing a signal that can be evaluated as an indicator of knock/no knock conditions. Note that these sensors are applied to provide a discrete indication of knock as opposed to an actual measurement of the knock phenomenon. They are normally applied in production, when the engine control system has already been calibrated, and the knock borderline operational limits measured and stored in an appropriate map or lookup dataset.

Note that structure-bourne detection of knock can be used for measurement. This is a simplified approach, particularly with respect to the sensors, but the setup of the system needs careful calibration, and the success rate of objective, repeatable knock measurement at the borderline of onset is limited.

Other techniques for knock measurement make use of optical access and ionization current. Optical access into the cylinder is becoming a more widely used technique thanks to advances in technology that increase functionality and reduce costs over time. High-speed photography and laser measurement techniques are still expensive and are confined to the laboratory because of the scientific nature of the instrumentation. Tomographic analysis is an expensive technique that remains time-consuming because of the extensive modifications needed to the engine in order to accommodate the measurement head gasket.

Ionization current has been successfully used for some time to establish knock conditions. Engines fitted with direct ignition systems can use this technique for what is basically a cylinder pressure measurement to distinguish the occurrence of the high-frequency knock components. This is a production environment technique wherein durability is more important that absolute accuracy; again, the engine controller is already calibrated with respect to the knock borderlines at operating conditions. Ionisation current is a possible technique for use in measurement environments and has the advantage that access into the cylinder for measurement is via the spark plug; thus, no intrusive installations or modifications are needed. The disadvantage is that ion current sensing requires a specific or modified ignition system that has the measurement capability integrated; this is expensive, and also means that the engine is not tested with the proposed ignition system, which could be a problem for some tests. Also, measurement of knock around the centre of the combustion chamber can be insensitive to circumferential modes of knock; these are the modes that have the highest amplitude and that can thus provide a signal where objective knock establishment is more reliable. The centric position detects the weaker radial modes, which can mean that the calibration of the knock detection threshold is more difficult to establish objectively. Note that it is also possible to instrument a cylinder with multiple ion probes for pressure wave and flame front growth analysis during fundamental research.

An alternative reduced-cost optical technology has become available recently that allows an optical insight into the cylinder via an instrumented spark plug. This allows measurement of flame growth in the cylinder; from this information, prediction of the probability of knock and its position in the combustion chamber can be measured objectively. The equipment consists of a measurement system that includes an instrumented spark plug complete with a multichannel optical transmitter and receiver unit. This is used in conjunction with appropriate analysis software for measurement, visualisation, and analysis. The most interesting aspects of this technology, which facilitate its use in more general applications, are

- Low cost when compared to other optical techniques
- Minimal intrusion: adapted spark plugs mean no modification to the engine, saving time and costs during setup for the measurement

- Integration with pressure measurement: certain manufacturers have fully integrated this measurement technology into the user interface of the combustion measurement chain such that a synchronous measurement and analysis of pressure, plus optical data, can be executed

For detailed studies of knock in combustion engines, the most commonly applied technique is via the combustion pressure measurement using the piezoelectric measurement chain. This produces repeatable and reliable results, and various algorithms are available that have been proven to give consistent determination of the onset of knock, as well as of the intensity and energy of the knock. This technique will now be discussed in more detail.

8.1.4.2 Knock Detection via Cylinder Pressure Measurements

Introduction

Knock is one of the most important abnormal combustion phenomenon. We have already discussed what knock is and why it occurs; generally, we can say that it is uncontrolled ignition of the gases in the cylinder, whether before or after the flame front. This auto-ignition causes the end gases to burn very rapidly, releasing energy at a rate much greater than would normally occur with controlled combustion—approximately 5–20 times the rate associated with normal combustion. The rapid energy release of high-pressure oscillations inside the combustion chamber produces the characteristic audible noise associated with knock. Figure 8.3 shows a typical cylinder pressure trace that is exhibiting the pressure oscillations due to the knock phenomena. It is important to note that the oscillations generally begin very near the location

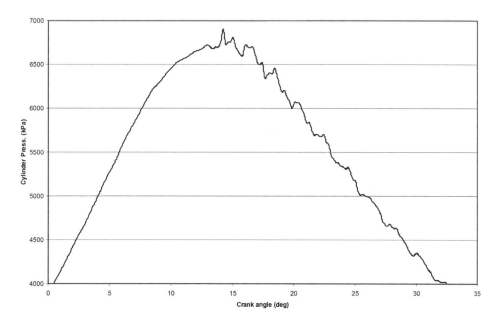

Figure 8.3 Cylinder pressure curve showing level of knock.

(Source: FEV.)

of peak pressure. Knock intensity can range from a very light, practically inaudible level to a very severe level that causes significant damage to the engine components.

Frequency of Knock

One consideration for the analysis of the raw data curve for knock measurements with respect to frequency is this: what is the frequency at which combustion knock occurs?

This can be approximated with a reasonable level of accuracy by calculating the natural frequency of the bore. Many passenger car engines have a bore diameter that causes the primary knock mode to occur in the 5–8 kHz range.

The primary knock frequency can be approximated thus:

$F_p = 570/B$

where

F_p = primary knock frequency (kHz)

B = bore diameter (mm)

For example, an engine with an 80-mm bore has a primary knock frequency of 7.125 kHz. Note, though, that there is a consensus of opinion that attention should also be given to the second mode of knock; this falls in the range of 10–16 kHz. So for the above example, the second knock mode would be approximately 14 kHz, but note, it is not necessarily double the first mode. However, it is suitable for a simple approximation and is offered as such. The knock frequency range of interest is an important consideration for the correct set up of the combustion measurement acquisition system.

Signal Processing

The maximum amplitude of the pressure oscillation owing to knock is commonly used as the basic input to many knock detection algorithms. It can be defined as the maximum positive value of the high-frequency components of the pressure curve within the knock window. It is also known as the maximum knock over pressure. This value is a very useful indicator of the severity and intensity of the knock event. The reason for this is that it is generated by the spontaneous combustion of the end gases as they ignite rapidly, and because the engine damage caused by knock results directly from the high gas pressure and temperatures in this end gas. These peak pressures also provide a meaningful, objective measure of the knock severity, and can be related easily to component loading and failure.

Filtering

Most combustion measurement systems use software-specific features for the measurement and analysis of knock. The cylinder pressure raw data curve is acquired via the usual measurement chain components. The signal is acquired and digitised by the system for analysis and storage. At this input stage, a filter is often applied to remove unwanted high-frequency components. Typically, a low-pass filter is used to remove undesirable high frequency noise that can be superimposed on the raw signals; this high frequency noise may come from a number of sources:

- Electrical noise from the ignition system or other electrical devices
- Structure-bourne noise from the engine mechanics (valve events, etc.)
- Resonant noise from the transducer caused by excitation by the knock frequency components

Most systems offer selectable filtering. Ideally no filter should be used, but in many cases this is not possible because of the presence of the above types of noise. A general rule is that the filter should be as high-value as possible to condition the signal appropriately for the measurement task. A typical value of 25 kHz is suitable for measurements in the normal operating speed range of a typical passenger car engines, in conjunction with a 0.1-degree crank angle sample rate.

The reason for this is that low-pass filtering has the undesirable effect of phase shifting the signal—shifting the relationship between the measured curve and the abscissa. The filter can then be considered a form of programmable delay: the lower the pass frequency of the filter, the longer the signal will be delayed.

This is normally not an issue for offline data processing, but for online or real-time signal analysis, this can be a problem, particularly for a combustion measurement system wherein the relationship between pressure and crank angle is particularly important for the accuracy of certain calculated results (e.g., MEPs derived directly from the combustion data).

Acquisition Frequency

The acquisition frequency is required to ensure that the raw curve is sampled at a high enough rate that none of the high-frequency components are lost due to aliasing or undersampling of the curve. The most important factor is the resolution of the angle encoder and the measurement table (i.e., resolution) for the input channel—that is, the cylinder pressure curve.

The frequency components of interest on the raw, measured curve are at a relatively high frequency (typically 10–20 kHz) and thus must be sampled at an appropriate resolution. There are numerous sampling theorems that dictate the measured frequency needed to prevent loss of detail, depending on the highest measured frequency component (Nyquist, Shannon, etc.). In addition, there are rules of thumb that are gained by experience, but in general, for effective discretisation of the cylinder pressure curve with knock components, the curve must be sampled at a frequency of least two times the highest frequency component—up to 10 times, if possible. Remember, however, that the acquisition frequency of the combustion measurement system is not constant in rate; because the data sampling is triggered via the angle pulses produced by the encoder, the acquisition frequency triggering is dependent on the engine speed and the encoder resolution.

For example, if we assume that the highest knock frequency components will be less than 20 kHz, then the minimum sampling rate should be 40 kHz, according to Nyquist. The minimum engine speed dictates the lowest sampling frequency that will be encountered: if we assume 1,000 rpm, then this dictates a sampling resolution of the signal at 0.15 degrees crank angle—

thus, a typical standard setting of 0.1 degrees will be more than sufficient. Clearly, this is the worst case; as engine speed increases, so does the acquisition frequency. This is not really a problem until the upper speed limit of sampling is encountered. Combustion measurement system have an ADC (analogue-to-digital converter) that digitises the incoming raw data curve on each input channel at a frequency dictated by the angle encoder marks (and engine speed); this has a maximum sampling speed capability that at a certain acquisition resolution will dictate the maximum engine speed. Current generation combustion measurement equipment generally have one ADC per channel that operates in the range 800 kHz–1 mHz. Because this is generally more than enough speed for most applications, the limit need not be considered; the issue becomes prevalent only when older equipment is used—particularly, previous-generation equipment that often shared ADCs across a number of channels. In such a case, one channel at high resolution would mean a certain engine upper speed limit; if two channels were measured, then that upper speed limit would be halved due to the sharing of ADC processing resource across the two channels, which increases the required conversion rate and reduces the maximum engine speed by a factor of 2. Note that high-resolution sampling also consumes memory in the same way; this must also be considered when setting up the combustion measurement system.

As a general rule, sampling for knock measurement and analysis is mostly recommended to be taken at 0.1 or 0.2 degrees crank angle in the knock window.

Acquisition Window

In addition to the filtering of the raw data curve, the dataset is often windowed, which means that it is acquired only within a specific range on the pressure curve where knock typically occurs. The data outside the window range will then be ignored, or acquired at a lower crank degree resolution, in order to optimise the available memory on the acquisition hardware. A typical range for this knock window is from the position of peak pressure to around 70 degrees after TDC. This range can be verified by initial measurement and analysis of pressure data from the engine to be tested; this is a typical task in the procedure of calibrating the knock detection feature of the combustion measurement hardware. Most combustion measurement systems generally support this windowing of the data for knock analysis. A multilevel measurement resolution allows the measurement of knock (when present) but also still allows the calculation of standard combustion results (IMEP, 50% conversion angle) without collection of large, high-resolution cyclic data sets that will consume memory unnecessarily and be difficult to handle in a postprocessing environment. This feature is essential if the combustion measurement system is to be used in a gasoline engine calibration environment.

Calculation of Knock Overpressure

Once the raw data is acquired in the correct range and resolution, high-frequency noise is filtered out, and the knock overpressure/peak value is calculated, as follows.

1. The measured curve is filtered via a moving average filter, producing a suitably smoothed version of the original curve that is then subtracted from the original, unfiltered curve (Figure 8.4).

This then produces a deviation curve that highlights the pressure peaks versus the crank angle, as shown in Figure 8.5.

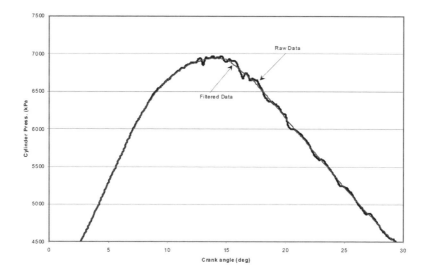

Figure 8.4 Filtered and unfiltered pressure curve with knock present.
(Source: FEV.)

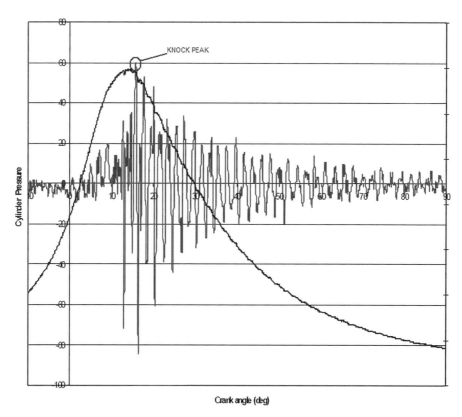

Figure 8.5 Knock peak overpressure curve.
(Source: FEV.)

From this curve, the peak value of the oscillations is extracted as a cyclic result. The filter type used to create the smoothed curve can be one of a number of types. A moving average filter has proven to give good results and is numerically and computationally efficient.

8.1.4.3 Further Processing and Calculation Methods

Although the knock pressure peak amplitude is the fundamental result that is often used for knock evaluation, once this basic result is derived, there are many additional further analyses and calculations that are performed in order improve the reliability and repeatability of the measurement. In addition, these alternative approaches with their differing analysis methodologies may be more suited to specific engine types or combustion technologies. The overall goal for the manufacturer of the combustion measurement system is to produce a technique for processing the cylinder pressure signal that is easy for the user to set up, that produces quality accurate and repeatable results, that responds quickly to the knock event, and that is not too excessively intensive with respect to calculation. (The latter is particularly important for real time execution.)

Knock evaluation is a subjective topic about which experts and manufacturers of engines and measurement equipment seem to have a number of opinions, particularly with respect to the best methodology. A number of proposed analysis methods are available using combustion measurement equipment currently on the market. In addition, many systems offer a degree of flexibility, allowing users to write their own algorithms or calculations for execution online during the measurement. This means that users of combustion measurement equipment may employ a number of approaches when analyzing the raw data curve.

Below are listed the most common approaches (at time of writing) for a number of established equipment manufacturers.

Knock Pressure Peak and Knock Intensity

This is a development of the methodology mentioned above: an extension to the knock pressure peak evaluation. The process involves rectification of the peak curve so that it has positive values only. This rectified curve is then integrated in the measurement window, producing an integral curve.

From these two curves, shown in Fig. 8.6, two result values are derived: the peak value (from the rectified curve) and the integral value (from the integrated curve). The two results are representative of the intensity (peak) and energy (integral) of the knock event in that cycle.

AVL Real Time

It is from these two basic results that the AVL real-time knock detection method evaluates whether a knock event has occurred. The basic process is shown in the flowchart shown in Figure 8.7.

The key to this evaluation method is the calculation of a base noise level. This is a dynamic calculation of a reference level cycle that takes into account background noise at that specific engine operating condition. In order to do this, the results from a number of engine cycles

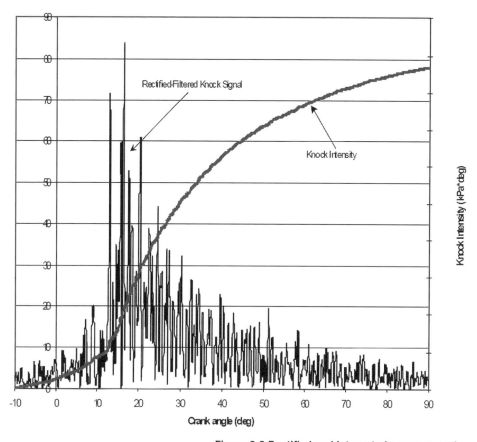

Figure 8.6 Rectified and integrated pressure peaks.
(Source: FEV.)

are averaged to produce reference peak and integral results against which the results from the actual cycle are compared. This means that noise and vibration from the engine that may affect the measured curve are effectively cancelled out for the evaluation (as this noise would affect the averaged values as well as the actual values). The advantage is that the base noise level is dynamic, and shifts as the engine operating point changes. In addition, the number of cycles for the base noise level calculation is selectable by the user during parameterisation; thus, the base noise level sensitivity can be adjusted with respect to the test conditions—that is, the number of sample cycles for base noise determination can be reduced to facilitate a more reactive evaluation (for example, during transient tests).

The peak and integral results are compared as a ratio to the base noise value. The value of this ratio is set by the user, and is a parameter used to tune the response of the system. For example, if the actual peak value is greater than the base noise level by a factor of 1.5, then a discrete value is set that can be used in the final evaluation. A similar approach is used for analysis of the integral value, thus giving two values for the knock analysis and event decision by the system, as a decision on knock criteria.

Figure 8.7 Real-time knock detection algorithm.
(Source: AVL.)

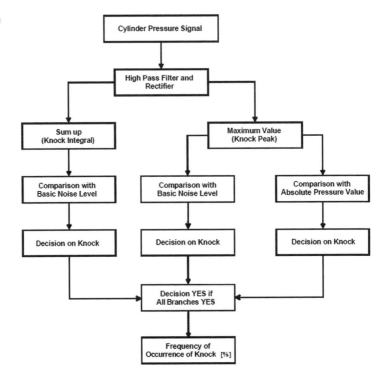

In addition, the peak value of the knock overpressure is monitored and compared to an absolute limit set by the user. The engine components have a operating limit with respect to pressure loading; intense knock peaks with high amplitude can cause considerable damage in very few engine cycles. This limit is generally speed-dependent. There are two threshold values that are speed-related (lower and upper), with linear interpolation between them. The upper or lower threshold value applies to rpm values outside the threshold points; a knock peak maximum under these thresholds is always ignored. The current threshold value is interpolated between the two rpm values; speed-dependent oscillation maximums with small amplitudes can be attenuated, if required.

The combination of these three values—knock peak ratio, knock integral ratio, and knock peak absolute—are used together with AND logic in order for the system to decide whether knock has occurred for a given cycle. A minimum of two out of the three criteria must be selected by the user in the parameters. These discrete values are then used to decide and record whether a knock event occurred. This is a result with a binary value relating to knock or no-knock condition for that cycle. This value can then be used as a trigger, via an external output, to signal that knock is occurring. This can be used as an alarm signal for the engine or test controller to react to prevent engine damage. The knock/no knock discrete signal is then statistically processed to produce an additional knock-related result. This frequency-related result given in events facilitates the knock frequency displayed in percent (%) of the last N cycles, where N is a evaluation window set by the user. As an alternative, the result can be a values of knock events per second (1/sec). The knock events per second are determined from the knock frequency in percent over the current engine speed.

In summary, the AVL real-time algorithm produces the following results available for calculation, display, and storage in real time:

- Knock pressure peak (value in bar)
- Knock pressure integral (value in bar × deg)
- Knock event (binary value)
- Knock frequency (%, or events/second)

Note that this method, like most knock detection methods, needs tuning to match the engine. It is definitely not plug-and-play. In order to achieve objective and reliable measurements the systems needs correct setting of the various parameters. It is advisable that once the system is operational in the test environment, several trial measurements be made at a number of speed and load conditions, running the engine into borderline knock (established audibly). While this is done, the resulting values of peak and integral should be monitored. In addition, the result of knock event can be calibrated to trigger at the appropriate level of knock intensity. Another setting to be observed and adjusted is the number of cycles for base noise level calculation; this can be set according the type of engine test. For steady state engine calibration procedures, 50 engine cycles will give a stable and repeatable response, but under transient conditions, the base noise calculation will lag; this may cause inappropriate system response. Under these conditions, the number of cycles should be reduced to around 20. Note that at low speed and light load, the pressure curves become smooth with no knock; consequently, the knock peak maximum value tends to 0, and in this case the dynamic base noise level detection will become too sensitive. For this reason, there are minimum thresholds set internally for the knock peak and integral values. Cycles with values below this are always considered as not knocking.

It should be noted that the installation of the transducers has an effect on the output signal, hence the requirement for tuning of the system response via experimentation. Any pipe oscillation will have an effect on the measured signal. It is recommended that if flush mounted transducers are not used, the knock integral calculation, being undependable due to oscillation frequencies, be used in conjunction with the knock peak pressure value.

The AVL real-time algorithm is well proven and reliable, but the disadvantage of this method is that setting the system up and tuning it can be time-consuming. In addition, the transient response is limited because of the requirement for establishing the base noise level. Certain more recently developed algorithms have better performance in this respect.

FEV CAS

The FEV approach uses the same basic information derived from the pressure curve: a peak and integral result value from each cycle. The peak value (known as the knock amplitude) is used as a basis for the system's internal knock meter calculation, in conjunction with a weighting table. The user of the system can adjust both the knock amplitude range and the weighting factors of the table per application, and to tune system response.

The knock meter function provides an objective method by which the level of knock can be characterised and measured, producing a value for knock based on the magnitude of

Weighting Factor	Minimum Knock Amplitude (bar)	Maximum Knock Amplitude (bar)
0.0	0	1
0.7	1	2
2.1	2	4
4.9	4	8
7.7	8	12
10.5	12	16
13.3	16	20
16.1	20	24
18.9	24	28
21.7	28	32
24.5	32	36
27.3	36	40
30.1	40	44
32.9	44	48
35.7	48	52
38.5	52	56
41.3	56	60
44.1	60	100

Figure 8.8 Knock meter weighting table.
(Source: FEV.)

the knock amplitude. As the knock amplitude increases, the weighting factor and value generated by the knock meter also increase. The knock amplitude (or knock peak) used in the calculation is roughly half the measured knock peak-to-peak value. The result value generated by the system from a single engine cycle is cross-referenced to the weighting table and the correct weighting factor is selected and applied, producing the final result value for analysis and storage.

Figure 8.8 shows the ranges for the knock amplitude and the associated weighting factors set at the system default. The maximum and minimum knock amplitude columns establish the classification range for application of the appropriate weighting factor. The range and weighting factor can be modified by the user to calibrate the system in order to develop the most meaningful output value for the final application of the data.

In addition, the system allows calculation of a value referred to as knock intensity squared. This is determined by squaring the rectified-filtered knock trace, then integrating over the range of the knock window. This result value obtained from this is relatively small when no knock is present but grows exponentially when knock starts, even at very low levels of knock. It thus can be used as clear indicator of borderline knock. The signal curve from which the result is derived is shown in Figure 8.9.

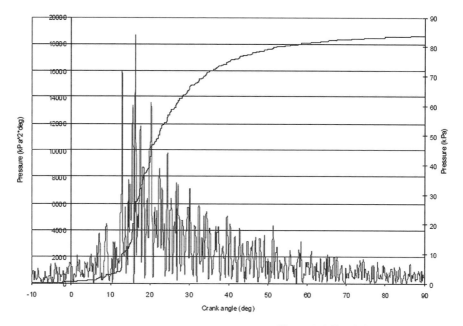

Figure 8.9 Knock intensity squared.
(Source: FEV.)

AVL KI (Knock Index)

The basis of the Knock Index algorithm is that low-level, borderline knock can be tolerated by the engine for extended periods without causing significant damage. It is the high-amplitude knock peaks that are most destructive; the knock detection algorithm should focus on the frequency of occurrence and amplitude of these events for the evaluation.

For the calculation, the knock pressure peak result (per cycle) is evaluated with respect to categories for the knock peaks. An amplitude range is defined, typically up to 64 bar, and is divided into categories for the knock peaks. Each category is given a weighting, which increases with the knock amplitude. A weighting is assigned to the result from each cycle, and the sum of all the weightings is calculated, and divided by the number of cycles.

The weighting table can normally be adjusted by the user to tune the response of the algorithm. For example, a small offset value could be applied to each result so that the low-value events are ignored with respect to the calculation. This method of knock evaluation is a statistical process, normally carried out over a number of engine cycles (typically between 300 to 1000, as this is shown to give reliable results). As a general rule, a KI value of less than 2.5 can be tolerated for extended periods without engine damage (i.e., borderline knock). An important point, however, is that due to the statistical nature of the calculation over a number of engine cycles, the transient response of the algorithm is relatively slow—for detection of knock for immediate reaction to prevent engine damage (within a few cycles), the amplitude of the knock peak (absolute limit) must be used as the alarm parameter.

AVL Transient

This is an alternative algorithm that is much more dynamic in response. In addition, it also has the advantage that only one cycle is needed for the evaluation of knock (i.e., only the current measured cycle is used); for this reason, the response during transient operation of the engine is superior to methods that need a prehistory of cycles (as described above).

For this algorithm, the cylinder pressure signal is first broken down into a low pass and a high pass. The position of the peak value of the low-pass portion is used to determine two evaluation window positions: one before the peak, and one after. Within each of the windows, the high-pass signal curve is integrated to give a characteristic result value determined from the rectified, window width–standardized integral of the high-frequency portion over the window range. The result derived prepeak is taken as the reference value, and is compared with the postpeak (knocking) value to produce a ratio. Using this method, the excitation of the interference from pipe oscillations, caused by the more pronounced pressure gradients shortly after the start of combustion, can be isolated from the actual knock oscillation. The ratio of the two characteristic values represents a measure of the knock intensity. If this ratio exceeds a user-definable threshold, the cycle is defined as knocking. The yes/no knock result for each cycle can then be used to determine the knock frequency; if a predefined frequency is exceeded, this could be used as a control signal.

The evaluation method is shown diagrammatically in Figure 8.10.

AVL Histogram

This method is used to evaluate knock behaviour specifically for use in engine calibration applications. The algorithm, like others, uses the knock pressure peak value as the basic input signal. From the knock peaks of each cylinder, a histogram is created in a moving

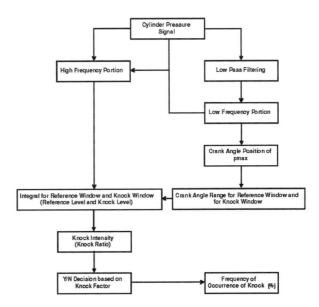

Figure 8.10 Transient knock detection algorithm.
(Source: AVL.)

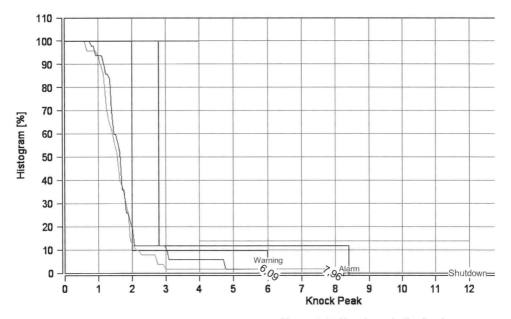

Figure 8.11 Knock peak distribution curves.
(Source: AVL.)

window—that is, a knock peak distribution curve. The normalized frequency (0–100%) of current knock peak occurrence is represented over the x-axis knock peak in the moving window. This produces a characteristic curve of knock peak distribution, observable during the measurement. An example is shown in Figure 8.11.

This histogram curve is checked cycle by cycle against defined limit curves (also shown) to produce index values for each cylinder identifying the knock severity for that cycle (if knock has occurred). These index values are classified according as follows, with the knock histogram index (KHI) assuming one of four status values:

0 = No knock

1 = Warning: Slight knocking. Typically, as a reaction to this level, the ignition would be adjusted in small increments during automatic engine optimization.

2 = Alarm: Heavy knocking. As a reaction to this level of knock, the ignition would be adjusted in large increments during automatic engine optimization.

3 = Shutdown: Dangerous knocking. In this case, the test would be stopped would be stopped in order to protect the engine.

During operation, if the histogram curve crosses the limit curves at any point, then the status of that cylinder is changed to one of the above values (0,1,2 or 3), according to knock severity. These cylinder and cycle specific results can be evaluated, stored, or transferred to other measurement or control systems for reaction.

235

The important aspect of this evaluation is correct setting of the limit curves for the three states that classify the knock severity. Each limit curve is defined by two corner points: the lower knock peak threshold on the x-axis and the related histogram limit on the y-axis define the limit curve's first inner corner point. The second corner point is defined by an upper knock peak limit. If, independent of the current frequency, a knock peak exceeds this upper limit, the relevant engine cycle will be considered to be knocking. For example, in Figure 8.11, the warning limit curve is defined by its pressure range on the x-axis (2–6 bar) and the frequency within this range on the y-axis (10%).

For setting the limit curves, only the warning limit curve needs to be adjusted by the user; the alarm and shutdown curves are derived by factors from the warning curve, which assists the user in simplifying the parameterisation. For this reason, correct setting of the warning curve is an essential prerequisite before starting the measurement procedure; this can only be done with the actual engine running. The limit curve is then determined by observation while the engine is intentionally operated from nonknocking condition to knocking condition. Because knock peak amplitudes can fluctuate significantly depending on the engine speed, static limits for the limit curves are not suitable; dynamic limits, depending on the engine speed, must be applied. For this reason, the algorithm allows that the limit values are adjusted for three engine speed points (low, medium, and high speed—representative of the full speed range) at full load; then, linear interpolation is applied between the speed points and is used to adjust the limit curves dynamically, during the measurement, according to actual engine speed. When engine speed values occur that lie below the low speed point, the limits do not change. For high-speed levels exceeding the highest speed point, the upper limit is fixed and applied.

Third Derivative

This is an interesting method that can be applied for detection of knock when constraints exist with respect to data acquisition frequency. In general, we have already discussed that, because knock events are relatively high-frequency (with respect to the fundamental frequency of the pressure curve), the data must be acquired at a high enough sampling rate to ensure no loss of these frequency components due to undersampling. This can be a challenge for the measuring system with respect to throughput and memory storage capacity. There are several advantages to a knock detection method that can utilise lower-frequency data, apart from the demands on the measuring system; an algorithm that uses the general shape of the pressure curve as opposed to the high-frequency components means that effects due to transducer location and mounting position have a lesser impact on the measured data and the measurement chain setup.

The method uses a characteristic knock indicator that is derived from the third differential of the pressure curve in a window range where knocking may occur. It identifies knocking via the severity of the pressure change that is associated with knock. A knock event consists of uncontrolled combustion of the end gases prior to the main, controlled flame front. This produces an abrupt, localised pressure rise. In addition to this, the pressure decay is also considerably accelerated due to the temperature and heat transfer conditions. This produces a characteristic shaped curve that has a rapid positive-to-negative curve slope. Because the curvature of the measured signal can be quantified by the second derivative, this rapid change would be associated with a large, negative, third derivative of the pressure curve. This is used as the characteristic indicator to discriminate between knocking and nonknocking cycles.

The method is further refined by the careful use of filtering of the signal to reduce noise without altering significantly the form of the curve (by the use of a cubic spline fit differentiator). This approach helps ensure that the differentiation process, which is inherently noise-amplifying, does not affect the reliability of the evaluation. In addition, an important factor is that this method does not have excessive requirements with respect to computational effort; this means that it can be executed online during the measurement, if suitably coded in the measurement system software.

During development of this algorithm, experimentation showed that it can be used reliably to indicate the severity, in addition to the occurrence, of knock via the peak negative value of the third derivative. Various configurations of transducer installation and data acquisition frequency were used to develop the algorithm and prove its performance. Although the author is not aware of any current combustion measurement systems that offer this method of knock evaluation as standard, if it were to be employed, it would have to be coded by the user in the measurement system open interface.

An important, possible application for this method of knock quantification could be for use in engine management control systems. It is likely, based on current trends, that combustion engines will need to be equipped in production with cylinder pressure sensing as standard. This will provide an input to the engine control system on the combustion event itself. For this application, the low computational power and low-frequency acquisition requirement make this algorithm very attractive. In addition, its insensitivity to high-frequency noise induced via electrical or mechanical interference means it would be quite practical for this in-vehicle application.

Further development of this method is likely—it is an interesting alternative to the other methods mentioned, all of which require higher-frequency data acquisition; this is more likely to occur in a research and development environment.

8.1.4.4 Considerations for Knock Measurements

Although there are a number of evaluation methods for interpreting and measuring knock with a combustion measurement system, there are a number of basic rules to follow, and observations to heed, with respect to the equipment set up to guarantee the quality of the measured data and the security and reliability of the measurement.

Transducer Position, Type, and Properties

The transducer is one of the most critical elements in the measuring chain for knock detection. Its properties and mounting position have a decisive effect on the reliability of the measurement. In the first place, the correct equipment must be chosen. The transducer itself must be capable of withstanding the extreme pressures and temperature that occur during knock. Under these operating conditions, the transducer has to give consistently reliable and predictable data. The transducer itself must have a sufficiently high natural frequency

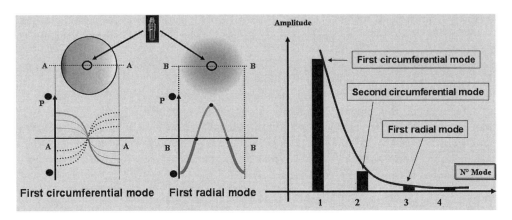

Figure 8.12 Modes of knock.
(Source: AVL.)

to prevent the effects of knock induced resonance's being superimposed onto the measured data, giving inaccurate results. For this reason, small uncooled transducers are favoured, as they generally have a high resonance frequency of above 100 kHz; they also have the advantage of being easier to accommodate.

Another consideration for the transducer is mounting position. This can have a dramatic effect on the sensitivity of the transducer to knock due to the various circumferential and radial modes of the knock pressure waves. Centrally mounted transducers (for example, mounted in a spark plug) are much less sensitive to the higher amplitude circumferential (or basic) modes; although they are much more responsive to the radial modes, these are of lower amplitude and interest, as they are not the main knock vibration modes. The knock modes are illustrated in Figure 8.12.

The front face position of the transducer is also to be considered. A recessed mounting creates a pipe or channel that can suffer from oscillation effects at certain frequencies (depending on the length of the pipe, gas temperature, and so forth). If these oscillations are near the frequencies of interest for knock, the measurement is no longer reliable and objective. For this reason, flush-mounted transducers are recommended to avoid resonance from this effect. One issue, however, is that flush mounting means that the measuring diaphragm is exposed in full to the pressure and temperature extremes; for this reason, transducer life can be expected to be reduced. This can be counteracted, to some extent, by the use of thermoprotection shields, as shown in Figure 8.13. These small perforated discs fit onto the front face of the transducer, thus protecting the measurement diaphragm from the direct heat of the flame front. They are normally employed in conjunction with small, uncooled transducers and can extend the life considerably when used in harsh measurement environments (such as those with knocking combustion). The reduced impact of the high-temperature blast from the flame front on the measurement diaphragm of the transducer has the additional benefit of improving the cyclic temperature drift characteristic by up to 50%.

Figure 8.13 Transducer thermoprotection for the transducer measurement diaphragm.
(Source: AVL.)

Measurement Range and Resolution

The sampling frequency of a combustion measurement system is directly related to the engine speed. For this reason, correct measurement resolution is critical to ensure that data is sampled without aliasing, at all the required engine speeds during the test. The sampling frequency of the measured data has an effect on the subsequently calculated result and thus can affect the reliability of the knock detection and calculated knock intensities. As a general rule, for most applications the data should be sampled at 0.2 or 0.1 degrees' crank angle within the knock window.

Setting of the knock window range is also an important aspect, as it is not practical to sample the full cycle at the resolution mentioned above, since this creates extremely large data raw sets that are difficult to store. Most combustion measurement systems support a measurement window within the main measurement range, and for knock applications this should be set appropriately to capture the knock event under all operating conditions of load/spark timing. Generally, peak knocking pressures occur in the range of 10 to 30 degrees' crank angle; commonly used ranges are 0 to 40 degrees' crank angle or 10 to 70 degrees' crank angle.

Measurement System Setup

Knock measurement data sets are generally quite large due to the sampling resolution, so the measurement system must be equipped with sufficient onboard memory to capture a reasonable number of engine cycles for an objective analysis. This is particularly important when a statistical calculation will be used to evaluate the number of pressure peaks, classified over a number of measured cycles (such as with knock index calculation). Often these statistical approaches are favoured when fast, real-time response is not required, and when repeatability and reliability of the knock evaluation is of greatest importance. This is because peak knocking pressure data has considerable variability; in these cases, the number of cycles needed to be sampled is quite large—several hundred, if not a thousand, cycles.

Another important aspect to be considered with respect to the measurement device is the bit resolution of the measuring channel ADC (analogue-to-digital converter). The dynamic range of the ADC affects the digital conversion quality and, analogous to the measurement resolution, if the number of available steps is insufficient in the measurement range employed at the ADC, the digital conversion will be low-quality, and the data will be ineffectively converted, with the risk of losing essential signal components or information.

Most commercially available combustion measurement systems now have 14-bit ADCs, many have 16-bit. The former should be considered as the minimum requirement for knock analysis. It is possible to use 12-bit ADCs when the raw signal should be prefiltered externally to remove the fundamental pressure curve frequency before applying to the ADC for conversion. This means that only the high-frequency components are sampled; thus, the requirement for the dynamic range is reduced. In this case, however, the data can only be used for knock analysis; no combustion-related results could be calculated due to the loss of the basic pressure curve component.

Optimisation of the ADC range is also important during the measurement setup for the same reason: good-quality discretisation of the raw measured curve. The pressure range of the measurement under the test conditions should be considered and the system set up such that as many of the available bits are utilised in the signal input to the input channel. For example, if most measurements are to be made at part load during a particular test, then the range setting of the amplifier charge to voltage transfer function must be sufficient to use as much of the input range of the combustion measurement system as possible. If then full load measurements are to be made, the required range should be approximated, and the amplifier adjusted to reduce the gain. Use of a single amplifier gain allows full load measurements without saturation of the ADC (i.e., over-range), but the bit utilisation at part load may produce poor-quality data after digital conversion.

8.1.4.5 Summary

The knock measuring function is an important feature that is available with many modern combustion measurement systems. For gasoline engine measurements it could almost be considered a prerequisite feature for measurement and analysis, in addition to providing engine protection. The above information gives an insight into a number of different techniques for knock measurement employed by the equipment manufacturers. To some extent, each one has its own ideas about and specific approaches to quantification of the knock phenomena. But most do support, via open software features, the possibility of the user's programming specific algorithms that can be written and executed during or after the measurement.

The most important factors for successful knock measurements are the fundamentals: correct choice and installation of the pressure transducer in the engine, and factors that affect the digitisation quality of the data (knock window range and resolution, optimisation of ADC input range). Simply, it can be stated that if the raw pressure signal is acquired efficiently, without extraneous noise or unwanted effects—and if this data is then digitised and stored with minimal or no losses of important signal components—the actual algorithm selected for the final analysis is at the user's preference, and is not the main factor for success of the knock measurement and analysis.

It should also be noted that in most cases, any knock algorithm needs some tuning and adaptation to the specific application before it can be employed for the actual measuring task. This is particularly important for engine calibration applications in which the reliability and repeatability of the knock measurement is an essential component in the overall test environment and facility. The knock measurement performance in this application is executed in real time, and during experimental procedures, engine knock borderlines have to be established accurately during the measurement, as these limits will dictate the steps of the test procedure and possibly have an impact on the success or failure of the overall test. For these reasons, it is prudent to spend some experimental time, outside the main test time, to set up the system, running some tests at various speed/load conditions to understand and the adjust the response of the chosen algorithm running on the combustion measurement equipment. This time can be considered well spent, as it ensures the reliability of the combustion measurement system performance during knock measurement and evaluation tasks.

The methods mentioned above are a small selection of possible methods for knock analysis chosen by the author, many supplied by manufacturers of equipment. But they should not be considered exhaustive. There are many excellent papers available from academic and practical work and experimentation that describe knock in much more detail, as well as the issues that are encountered in real-life applications. It is strongly recommended that this information be sought in order to gain a more detailed understanding of knock that goes beyond the scope of this book. Once this has been gained, it will then be possible to make an objective, informed decision as to which algorithm is correct to employ for a specific application. This is an important point: there is much choice among the various methods because of their suitability for different engines and operating applications.

There has been much experimental work done to understand the knock phenomena, and research is ongoing to try to understand knock with respect to modern engines and new combustion technologies. Engine knock is one of the main limiting factors that prevents further improvements in the cycle efficiency of spark ignition engines, and thus it is a major challenge in the engine engineering and scientific community as they try to further understand the reasons for knock, and how to overcome the limitations surrounding it.

8.2 Combustion Noise

8.2.1 Introduction

Reducing vehicle noise emissions is a topic of great interest to vehicle manufacturers. The more that the noise radiated in and by the vehicle can be reduced, the more comfortable the passengers in the vehicle are, and the more disturbance to the environment is reduced. Vehicle noise measurement is a specialised field that generally involves measurement of the airborne noise as the vehicle passes by. It involves a specific test environment, for the complete vehicle—which is costly to accommodate. Anechoic cells for engines and vehicles are often used, but, again, these are specialised facilities with specific noise measurement equipment and skilled staff for the measurement execution and data evaluation.

Often, the engine engineer will be required to measure and understand the noise contribution from the engine alone. Of course, this noise consists of a number of elements combined—that is, mechanical noise from the engine parts, noise from the combustion event itself, induction, and exhaust noise. For the engine engineer or developer, particularly with respect to diesel engines, the noise that is generated by the combustion event alone is of most interest. Modern diesel engines have electronic control systems with high degrees for freedom allowing for the adjustment of many of the engine parameters during the calibration process, and these have to be optimised to give the best compromise between noise, emissions, and fuel consumption. This is because the rate of combustion can be controlled via the injection strategy, which has an impact on the noisiness, and perceived smoothness, of the engine. In an engine test bed environment, these parameters have to be optimised in a parallel approach in order to achieve the best compromise of performance and economy. In addition, the engine smoothness, which in turn contributes to overall drivability, is now considered important as an optimisation target.

The most efficient method for evaluating noise from combustion is via the pressure signal itself; the development of the noise from combustion is a direct result of the pressure rise in the combustion chamber. It is thus a measurement task that can clearly be executed via a combustion measurement system, and it is an often required function needed for the development of modern diesel engines. The combustion noise evaluation method uses a frequency analysis of the cylinder pressure curve, so the noise value returned is specifically associated with the combustion. Other noise contributing elements from the engine structure or ambient conditions are therefore excluded and do not influence the result, thus making it ideal for an objective evaluation during the development phase.

8.2.2 What Is Combustion Noise?

With respect to combustion measurement, the combustion noise is the noise radiated by the surfaces of the engine structure as they vibrate and resonate in response to the rapid development of pressure in the cylinder during the working engine cycle. The high pressure developed in the cylinder deflects the engine structure, causing it to resonate in a number of vibration modes. The external surfaces of the engine that are in direct contact with the surrounding environment radiate this energy as sound pressure.

In simple terms, this noise can be directly linked to the combustion event and the rate of pressure rise. However, it is not possible to derive the combustion noise directly from the pressure rise; experimentation has shown that the correlation is poor and cannot be relied upon for all engines and operating conditions. Therefore, alternative methods of deriving the combustion noise have been developed; these involve the measurement of cylinder pressure data, with subsequent filtering and Fourier analysis. After this process, a transfer function is applied that represents the structural response of the engine. Establishing the properties of this transfer function, however, is not a simple process. The engine has to be operated in anechoic or semianechoic conditions in which the engine noise can be measured accurately. The engine is then operated in a condition (via timing or fuel adjustments) such that the combustion noise is dominant. By subtracting the cylinder pressure frequency levels from the measured sound frequency levels, the structure response can be derived. Figure 8.14 shows an example measurement.

Because well-documented experimentation has proven that the attenuation properties of most engine structures are very similar, a standard, average structural response curve

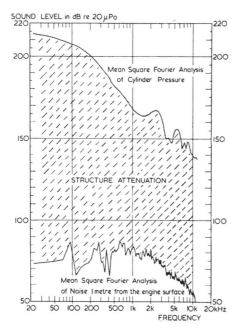

Figure 8.14 Structure attenuation derivation.
(Source: Delphi Diesel Systems.)

was proposed (often known as SA1). Further experimentation has validated this in that a mean, free-field structure response curve has been proposed and is generally accepted, and used, in combustion noise calculations by the manufacturers of combustion measurement equipment. This allows the calculation of a reasonably accurate but very repeatable result that can be obtained simply from the engine pressure data. An example of this curve is shown in Figure 8.15.

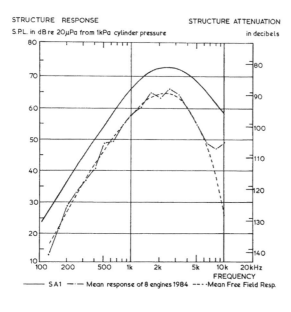

Figure 8.15 Mean free-field average structure response curve.
(Source: Delphi Diesel Systems.)

243

This measurement technique is useful to the engine developer because normally, in order to measure engine noise, a complex acoustic test environment is needed, as well as a subsequent process to separate the mechanical and combustion-related noise effects. If the engineer can measure combustion noise alone, the effects of adjustments or optimisation to the combustion system and related components on the noise produced can be evaluated during the development phase without the above expensive measurement techniques.

8.2.3 Measuring Combustion Noise

Signal Processing and Calculation

There are several manufacturers of combustion noise meters; in addition, manufacturers of combustion measurement systems can include the combustion noise evaluation in their system software. There are some variations in the approach to the calculation, but in general, the method is as follows:

- Prefiltering of pressure curve data
- Application of a filter to represent the attenuation of a standard engine structure
- Application of a filter that represents the human ear response
- Calculation of total noise in dBA as a result for display or storage

This process is shown in Figure 8.16.

For signal acquisition, the normal measuring chain components for signal conditioning are utilised: namely, the pressure transducer fitted in the combustion chamber and the charge amplifier. It is important to note that if the combustion noise measurement is made using a combustion analyser or similar equipment, a critical factor is good quality digitisation of the raw signal. Therefore, correct settings of the charge amplifier to ensure optimal utilisation of the analogue input channel range, and an appropriate resolution for the measurement acquisition frequency, are extremely important (analogous to knock measurement setup conditions). Stand-alone combustion noise meters are normally preconfigured and measure in the time domain only; no user adjustments are normally required.

Once the raw pressure curve is available to be applied to the input of the measurement device (combustion measurement system, noise meter, or data acquisition system), the actual signal processing method to return the noise value depends on whether the device is

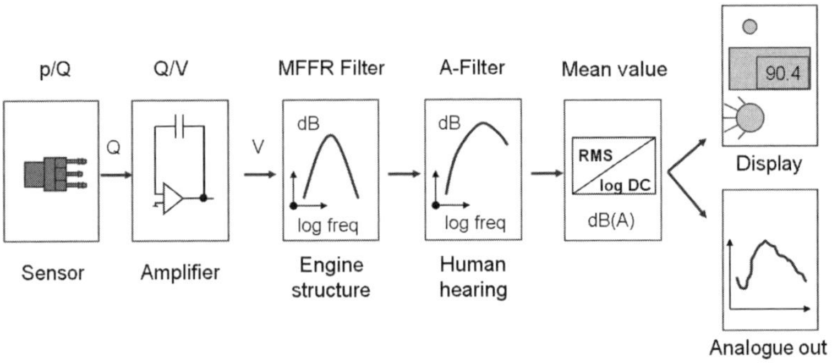

Figure 8.16 Method for combustion noise calculation. (Source: AVL.)

an older, analogue type, or whether the calculation is executed within a measurement device and is processed digitally. As a general rule of thumb, stand-alone noise meters are generally analogue, whereas combustion analysers normally use digital signal processing. Processing the signals digitally has a number of advantages: among them, behaviour of the filters is closer to ideal, and there is no drift. In addition, digital devices tend to be more flexible when adjustments or tuning of the filter responses are needed.

For digital processing system, which are generally combustion measurement systems, the data will be acquired in the angle domain; this requires conversion to time base before application of a Fourier transform (FFT = Fast Fourier Transform) to convert the signal to the frequency domain. Time base conversion requires measurement of the cycle time (less accurate) or the cumulative time difference between angle marks (more accurate). The Fourier transform produces a power level spectrum (referenced to 2×10^{-10} bar), which is then converted to a third-octave spectrum as shown in Figure 8.17.

The amplitudes of the spectral line at each third octave then define the mean signal level, in dB, within each third-octave band. From this third-octave dataset, the filters are applied (discussed below) in the frequency domain to modify the curve according to an engine structure and human ear response. Once the filtering has been executed, the third-octave bands are logarithmically summed to produce a single result, which represents the combustion noise contribution for that engine cycle.

As mentioned previously, combustion noise measuring meters are generally analogue devices that measure the pressure curve in the time domain. The raw pressure curve signal,

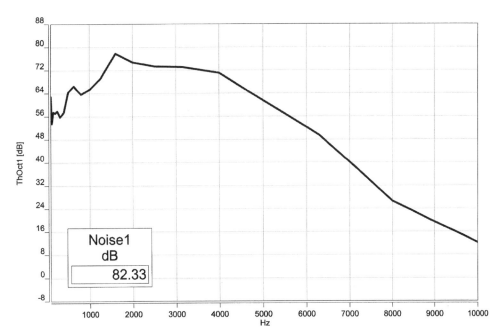

Figure 8.17 Third-octave spectrum.
(Source: AVL.)

output from the charge amplifier as an analogue voltage, is applied to the input of the device. The signal is filtered using analogue versions of the structure and human hearing response filters. The resulting filtered signal is then rectified via an RMS (Root Mean Square) circuit and applied to the output circuits via a logarithmic amplifier giving a result value in decibels (dB) referenced to 20 micropascals. Generally, the output of a stand-alone noise meter is a digital display wherewith results can be recorded manually, or is transmitted via an analogue output voltage wherewith the results can be recorded on another data logging device. (Note that the displayed result may not always be a real-time, cyclic value.) Such devices average the output over a number of cycles to give a statistic value that is updated regularly, which is more appropriate for steady state test conditions.

Filters Used in the Calculation

We have already discussed the structure response curve, a filter that replicates the attenuation properties of a standard engine structure via a frequency dependant weighting, and that is equivalent to the mean free field response. In most cases, this standard curve will be completely suitable to represent the structure of the engine under test, and it facilitates a realistic and repeatable noise measurement. This standard curve was derived via experimentation and comparison of many different engine types. During this process, it was found that the properties of most engine structures were remarkably similar in their response and that a single standard response curve could thus provide a representative result for most applications. Note, however, that the noise level derived is not an absolute measure of the combustion noise. Rather, it is a relative figure that enables the comparison of combustion system variations during the development process. The response of this engine structure attenuation curve is shown in Figure 8.18.

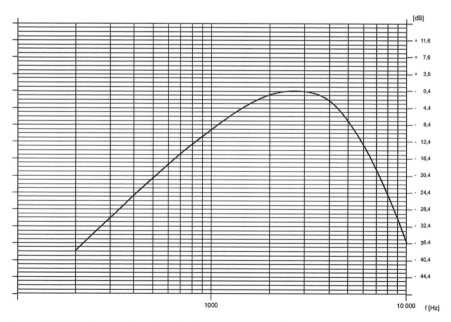

Figure 8.18 Standard engine structural response curve for combustion noise calculation.
(Source: AVL.)

Use of this curve in measurement and analysis equipment has been well established for a number of years. More recently, though, engine designs for the diesel engine have changed enough that this curve has become less representative. Engine structures are more sophisticated due to the advances in design techniques, and in addition, material technology has changed considerable since the engines that were compared to derive this curve were designed. This means that it may be necessary to tune this curve so that it represents more closely the properties of the actual unit under test. This structure response information for the test engine may be available from simulation code, or from acoustic measurements. This allows a response curve to be used that has been based on more realistic information to derive the structural attenuation of the actual test engine.

Many combustion measurement systems facilitate the option to adjust the weighting factors of this curve to match the actual requirements. This is relatively easy when digital filters have been employed in the calculation, as is common with modern combustion measurement systems. This is generally adjusted via the user interface, by adjustment of script coding or via a user dialogue. Because older combustion noise meters use analogue filters to represent this curve, changing or adapting of the filter curve is more difficult in this case due to the hardware component replacement needed.

The pressure curve information is modified by two filters to achieve the combustion noise value. One represents the engine structure; the other (known as the A filter) is used to represent the sensitivity of the human ear. The A-filter attenuates the low and very high frequencies to simulate the variation in hearing sensitivity with respect to frequency typical of human hearing. This curve is not normally adjustable; the measurement is made with this filter either on or (less commonly) off. The curve property is shown in Figure 8.19.

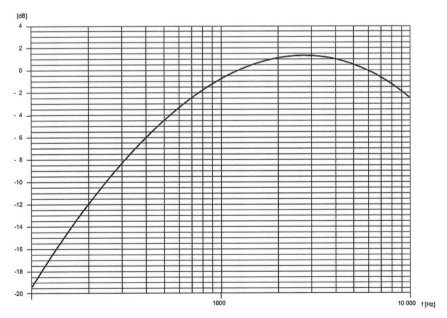

Figure 8.19 Filter curve property representing human ear response.
(Source: AVL.)

Additional filters may be applied in addition to the structure and the human ear response filters. Depending on the installation of the transducer and the quality of the signal produced, it may be necessary to filter out additional higher-frequency resonance from the combustion chamber in order to prevent these frequency components from affecting the noise calculation. Many combustion measurements systems allow for the application of adjustable digital filtering to signals before processing. For noise meters, often a selectable and range switchable low-pass filter prefilter is available if required. This low-pass filter must be applied carefully, as it can have a significant effect on the calculated noise result.

8.2.4 Summary

The ability to derive combustion noise directly from cylinder pressure data is a useful tool in the development of diesel engines and related fuel injection systems. Optimisation of engine noise is becoming increasingly important in the development process and is often a calibration strategy target for diesel engines in addition to performance, economy, and emissions. The noise measurement derived from the pressure curve facilitates an understanding of the noise from combustion without expensive acoustic measuring equipment or environment. It is therefore a logical functional extension for a combustion measurement system, in addition to being applied as a stand-alone device.

When the measurement is made via a combustion measurement system, the raw data is collected in the angle domain. This has an impact on the acquisition frequency (depending on the engine speed). For a good-quality evaluation when using this equipment, appropriate digitisation of the signal is essential, so amplifier range settings and data acquisition rate should be considered carefully when setting the system up prior to the measurement.

Also, if an environment or test field exists in which noise measurements are made using stand-alone analogue meters and digital combustion analysers, the noise results calculated will not be directly comparable; the noise calculation is a relative result for comparison, not an absolute value. The differences in analogue and digital processing mean that the calculated noise results could be different for the same engine at the same operating conditions with devices that use analogue or digital processing. For this reason, it is imperative that the same type, even the same device, be used throughout a test program.

Note that current requirements often need a specific attenuation curve for the engine structure, as we have discussed previously. This should also be managed carefully to ensure that the same structure response curve is always used in every test in which comparisons between tests and engines will be carried out. The same applies to any low-pass filtering that has been applied before calculation of the noise result. All filter properties must be the same for each measurement from which data is to be compared.

8.3 Cold Start

8.3.1 Introduction

Cold start, in the context of combustion measurement, generally refers to a measurement mode such that the start of the measurement is initiated from the first movement of the crankshaft—from zero speed. Because of this, the measurement of the engine startup process is often carried out on a cold engine (hence the name). This is significant, because normal measurements are started when the engine is running; often the hardware executes some synchronisation and phase checking before data acquisition starts. The consequence of this is that there is normally a slight delay between starting the measurement in the user interface and the actual start of data acquisition and storage. In most cases, with a running engine, this delay is not significant and does not impact the test. For certain engine start tests, though, losing cycles of data at the start of the measurement is not acceptable. Often, cold start test are exactly that, in order to understand the performance of the engine during the start and initial running phase at low temperature. This is particularly important for emission measurements, as the first few firing cycles contribute massively to the overall regulated emissions mass during a drive cycle. In addition, during engine calibration procedures at an engine test bed, the engine controllers must be optimised to ensure fast light-off of exhaust after treatment systems; in order to achieve this, an understanding of the in-cylinder processes during this phase is essential.

Another application for cold start is diagnostics—to understand the performance and reliability of the engine under extreme conditions with respect to starting and runup. In modern engines installed in vehicles, there is a particularly high expectation from the driver that the engine will start quickly under all operational conditions, with varying fuel quality or alternative fuels; this is a development challenge, and the more information engineers have about the combustion system operation during first movement of the crankshaft, the better they can optimise the system.

8.3.2 What Is a Cold Start Measurement?

The cold start measurement involves starting the data acquisition right from the first movement. This sounds a relatively simple task, but it should be remembered that the combustion measurement hardware normally incorporates features to ensure that the data collected is correctly synchronised with respect to TDC and phasing before acquisition and storage. In addition, it should be remembered that generally the encoder is not an absolute position-sensing type; thus, the absolute position of the crankshaft is not known from the first incoming encoder marks. In order to understand the technical challenge, let us consider how a normal measurement with a running engine would typically start:

- Initiate the measurement via user interface soft key or remote command.
- Hardware starts acquisition after receiving the first trigger mark.
- After third trigger mark, one complete cycle has been acquired; this is then checked for correct phase relationship.
- TDC offset is applied to the curve data, and the above phase shift is applied to correctly position the curve relative to the absolute TDC position.
- Now that the data is correctly aligned, measurement and storage can begin.

This process often takes a number of cycles. Although some combustion measurement systems allow bypassing of the phase correction to reduce time between initiation and measurement start and acquisition, some information is nevertheless always lost at the beginning of a measurement process during a normal measurement start. Another technical limitation that could affect this measurement task is that often it is required that the measurement start before the engine actually begins to rotate, in order to capture all the details of the combustion process prior to and during first movement. Because most systems, during a normal measurement, will not start measuring until trigger or encoder marks are received, synchronisation of the combustion measurement, with other measurement devices can be difficult and time-consuming.

For these reasons, cold start measurement mode has been developed for most combustion measurement systems available on the market. Typically, in this mode, the systems can support measurement start before movement of the engine or, alternatively, with the first crank degree marks from the encoder. Based on this, the data would be acquired in either the time or angle domain, or both. This mixed mode of data abscissa can often be required, as the combustion pressure data will be needed in the angle domain in order to calculate cyclic results as well as in the time domain for comparison and synchronisation of the data with other time-based measuring devices.

8.3.3 Typical System Configuration and Results

8.3.3.1 Preconditions prior to Measurement

Generally, the hardware interface connections for a cold start measurement are the same as for a standard measurement. Only certain older pieces of combustion measurement equipment need a specific configuration of the wiring to support this mode (particularly with respect to the encoder wiring).

With respect to the encoder, it should be noted that any multiplication of the encoder marks via hardware should be disabled, as this electronic mark interpolation does not generally work accurately at cranking speeds due to the large engine speed variations within the cycle.

Note also that the charge amplifiers for piezoelectric transducers must have drift compensation switched off. The reason is that the drift compensating process can distort the raw signal, as the cycle time is relatively long for cranking speed measurements. Bearing this in mind, the cleanliness of the leads and connections between the amplifier and transducer in the test environment is imperative in order to prevent charge leakage, so that an appropriate number of cycles can be recorded before drift effects are seen on the measured data. The consequence of this drift is that it could cause overranging of the combustion measurement system input channels and subsequent failure of the measurement task.

Also, the calibration of the pressure transducers should be considered; often transducer suppliers provide calibration information that is referenced to a specific operating temperature, and several calibrations will be offered according to this. For normal operation, the higher-temperature options would be chosen, but for cold start testing, the lower temperature calibration may be more suitable, depending on what part of the test is of greatest interest and needs the highest accuracy.

The dynamic input range for the signal into the measurement system that will be required during the test is a factor to be considered. The user of the combustion measurement system must take into account the maximum peak pressures likely to occur during the measurement. If this is known, then the amplifier setting should be optimised to ensure that the full input range of the measuring system channel (allowing for drift tolerance) is used to ensure good-quality digital conversion of the pressure curve at low speed, where peak pressures and signal range will be relatively low. This setting could be quite different from that used for a normal engine running measurement. Of course, it will depend on the engine operating conditions during the startup test duration; it may be necessary to use one amplifier gain setting for a startup-to-idle condition measurement, then change the setting for an idle-to-loaded measurement. This approach facilitates the use of settings, for the optimum conversion quality of the measured curves, that will be stored under the operating conditions of the engine during the test procedure. Based on the author's experience, it is less likely that the test's requirement will be to measure from cranking to full load conditions in a single measurement, but if this is necessary, this is a considerable challenge for the measurement system setup, as the dynamic range needed to efficiently digitise the raw data under all operating conditions could be greater than that offered by the combustion measurement system. This requirement is more likely if the combustion measurement equipment is fitted with an extended dynamic memory that facilitates the capture of a large number of raw data cycles (in the order of thousands of cycles), because then a complete start, warm-up, and loaded test cycle could be measured and stored without stopping the measurement procedure. In this case, external conditioning of the signal (prefiltering or differentiating) prior to digitisation at the combustion measurement device signal inputs may be required in order to accurately record the signal within the dynamic range of the combustion measurement system.

Another important prerequisite is that the correct TDC relationship has already been determined and stored in the parameter file. This is essential as the TDC offset information is applied retrospectively to the data acquired, immediately after measurement.

Obviously, as the data is recorded first, then corrected after the measurement, the calculation of results during measurement is not possible; hence real-time calculation or processing should also be disabled prior to the measurement task to conserve system resources.

Certain systems employ an internal timer/counter to the incoming encoder marks. According to the setting of this counter (i.e., frequency), its range (i.e., maximum bit count) and the angle encoder resolution dictate the maximum and minimum engine speeds—that is, the counter must be set so that it is fast enough to capture engine speed accurately, even at the upper end of the speed range, but it also has to be slow enough that the counter does not overrun at low speed. In most cases, the manufacturer setting is acceptable for most measurement conditions, but at cranking speed, the counter rate should be checked against the lowest engine speed (i.e., longest cycle time) to ensure that the counter will not overrun.

Also note that if the measurement will produce time-based data, a similar precheck of the acquisition frequency must be made to ensure that the data is sampled at a high enough frequency at the maximum encountered engine speed during the test to prevent aliasing of the raw signal curve detail.

8.3.3.2 Executing the Measurement

It is important that these preconditions be checked, and that the parameterisation is correct, before the start of a measurement, particularly if it is a true cold start test under low-temperature conditions. If the measurement fails due to incorrect settings, the vehicle must be reconditioned again back to the test temperature, which can waste hours of valuable time.

Assuming that the system is set up correctly, the measurement can be initiated via the software user interface or remote signal from a host controller (i.e., test bed control system). Once the measurement is started, the system then responds according to its operating mode. If the system is set to acquire an angle-based measurement, it will wait for a signal from the angle encoder. Once the engine turns and encoder marks are applied to the input, the system starts to acquire data as normal. Typically a test of this type will be a measurement over a number of cycles, depending on the memory depth of the combustion measurement device or according to the test pattern. The data will be displayed on the screen as acquired, but this is often meaningless, as the TDC relationship is incorrect during measuring unless it is corrected online by the combustion measurement system.

Once the required number of cycles has been measured and stored, the combustion measurement system typically applies the TDC setting of the loaded parameter file; this shifts the pressure versus volume table relationship for the whole measurement file, thus correcting the misalignment due to incorrect phasing so that the measured data curve is correctly adjusted for TDC. In addition, the system adjusts each cycle with the appropriate zero-level correction algorithm to provide a cyclic data set ready for calculation of the results of interest.

In summary, the data is acquired from the first encoder mark, and then the normal TDC and zero-level correction are applied after the measurement has been executed and the data stored.

If the measurement is to be time-based, once the system is initiated, acquisition is started immediately according to the predetermined frequency. This means that the measurement must be started promptly to ensure that the memory buffer is not wasted on low-value data prior to engine movement. As soon as the engine moves, the pressure curves are measured in the time domain as a single continuous curve, rather than a sequence of cyclic data sets. Often this data is valuable in order to examine the general shape and form of the pressure curves during initial firing; the disadvantage is that the fixed time-based acquisition frequency during the measurement means that at low speed the data could be oversampled (which wastes memory), and that there is a risk of undersampling once the engine has fired and is running at higher speeds. In other words, the fixed acquisition frequency will always be a compromise. In addition, no volume table is available as there is no crank degree mark, and thus no result calculations of the typical results of interest can be executed on the raw time-based data. The only exception is if encoder marks have been recorded in the time domain as a channel, in which case intensive postprocessing allows conversion of the data and subsequent calculations, although this is a very time-consuming exercise.

It has been mentioned that often, for this measurement task, the data may be useful in the time or angle domain, depending on the application. Certain measurement systems have software and hardware features that allow the measured data to be available with in both formats, either via sampling the same signal on two channels simultaneously (both a time-based and an angle-based channel), as it may be possible to allow channels (or groups of channels) to operate in either mode. Another alternative is that the encoder pulses are measured on a counter channel to determine the time difference between mark edges. This then allows conversion of the crank degrees into time by cumulatively expressing the time difference record between each crank angle mark. This creates a high-precision time base for the measured channel data, even though the sampling is not equidistantly spaced. Also, the system may be capable of producing two data sets for each measurement, one time-based and one angle-based. In such a case, the data is immediately converted from one domain to the other, and a new file is produced automatically by the system software after the procedural measurement.

8.3.3.3 Data Visualisation and Processing

The above hardware settings relating to system setup simply allow the data to be captured in the correct format. In simple terms, the system is just capturing, digitising, and storing the raw curve data sets in the combustion measurement hardware memory. The intelligent part of the process is executed by the operating software and user interface.

An efficient system will have software features that support the startup/cold start measurement and that allow fast correction and visualisation of the data. For example, once the measurement is finished and the data collected, the system should correctly apply zero-level and TDC offsets such that the data is correctly displayed, ready for further offline calculation of results or permanent storage. Often, in preparing the data, it is required that missing data points (i.e., from the start of the measurement) be populated with dummy values to give the correct shape to the displayed curves and prevent calculation errors caused by null values.

The user and graphical interface normally displays the curves from all cylinders overlaid but in continuous data curve format; this shows the in-cylinder pressure development during the cranking and then the first firing process. From this information, the engineer can evaluate the engine performance and successful firing probability of the engine. Figure 8.20 shows a typical display window with cold start measurement data. The data is shown in the time domain with a millisecond abscissa; each cylinder curve is offset and displayed in the correct firing order, and the black curve shows the instantaneous speed of the engine.

8.3.4 Summary

Cold start measurement is often required in the development of engines for vehicle applications, and specific functionality to support this measurement mode is a useful extension for the combustion measurement equipment in use. As discussed, there are certain preconditions and settings for this task that should be observed in order to support a quality measurement. This is particularly important when cold starting performance will be assessed. If the measurement fails, the vehicle has to be reconditioned again to low temperature conditions, which is expensive both in money and time. All efforts should therefore be concentrated on the reliability of the measurement system.

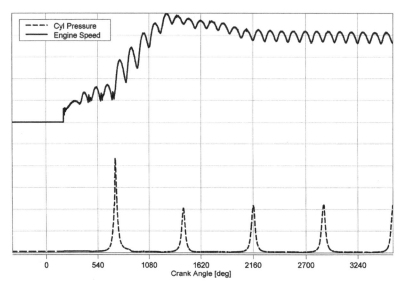

Figure 8.20 Typical display window of cold start data.
(Source: AVL.)

Once the measurement is made, the ability to visualise the information quickly is essential to plausibility checking, and to provide guidance in the next step of the test process. It is important that the combustion measurement system support this via software and hardware features and functions in addition to the possibility of displaying the data with alternative abscissa (time or angle).

The demands on the hardware in this mode are quite high with respect to required acquisition frequency and the depth of memory required to capture a reasonable number of cycles to measure the engine start process from first movement to stable idling operation. In addition, the system must handle correctly incomplete cycles, encoder errors caused by high instantaneous speed deviations, and the possibility of reverse rotation of the engine resulting from pressure hitback.

Chapter 9
Successful Measurements

9.1 Problems and Errors

9.1.1 Introduction

The combustion measurement system and its associated components that form the measurement chain are relatively specialised and have specific characteristics in relation to their applications. This means that many problems associated with the measurement can also be quite specific and therefore require a solid knowledge of the measurement application and the measurement system in order to resolve issues affecting the quality of the measurement.

Most errors associated with the measurement chain will cause one of two common symptoms. The first is that the quality of the measured signal, as observed, will be insufficient for storage or analysis. For example, excessive noise may be visible on the trace. This means that there is an opportunity for a skilled user to make changes or adjustments, to try to improve the situation. The second symptom will be less obvious—and perhaps only visible when the data is post-processed. For example, a calculated result may not be as expected. An error of this kind is more problematic because often, once the data has been collected and stored, the engine may have been removed from the test stand, meaning that observation of the test condition is no longer possible. The person responsible for data analysis has to make a judgment, based on the data, as to whether the error is due to the measurement chain or system setup, or whether they are observing a real measured effect that has occurred as a characteristic or function of the engine or test. Experienced users and experts in engine combustion can often detect errors in the data and make an informed judgment about their source. Following this, the data can be manipulated, if required, to offset the error. An example might be the TDC being incorrectly defined, giving error IMEP values. Experienced engineers can recognise this by knowing their engine and by close examination of the raw data and other results. From an educated guess, the data can be adjusted to give the correct IMEP, during processing and analysis. Then all results can be recalculated and checked. This is a time-consuming process, but at least it means that the data is useable and test-bed time is not wasted. The disadvantage is that an element of nonmeasured effect is introduced, which is then effectively not representative of the measurement conditions.

Generally a "right first time" approach is needed for combustion measurements. That is, check the setup twice, so that the measurement only needs to be taken once. A characteristic of this measurement, whatever system is used, is that only relatively few channels are measured (when compared to a typical test bed for research, where many pressure and temperature channels are often used, in addition to measurement devices and systems). But many calculations and results are derived from the raw data. Many of these results are sensitive to the quality of the raw data, and for this reason, small errors in raw data channels present large errors in calculated values—generally, greater by an order of magnitude. For this reason, setup of the system, pretesting, and then observation during the measurement are critical.

For inexperienced users, it is not always clear what to look for and what is significant. The following section attempts to highlight the main parts of the measuring chain and identify where problems can occur, providing some basic guidelines.

9.1.2. Typical Sources of Error

9.1.2.1 Transducer

The transducer is the source of the signal, and as such, is the greatest potential source of error on the measured signal. Though many of the errors can be eliminated by selection of an appropriate transducer and correct installation, the transducer operates in a harsh environment, and this can affect the pressure measurement and the quality of the signal from the transducer.

One of the main effects is drift. This is induced by the temperature changes that the transducer is exposed to. Drift is normally categorised into long-term drift (across cycles) or short-term drift (within a cycle). Drift has been discussed in an earlier chapter in this book. The main problem that drift causes is a shift in the sensitivity of the transducer that distorts the actual measured value. This change is brought about by the heat load acting on the transducer body and diaphragm, causing changes in the prestresses that the measuring element and its housing are subjected to. This causes a change in preload on the measuring element itself and consequently a change in its sensitivity. Note that it is less likely, with a modern transducer, that the measuring element itself will change sensitivity due to temperature change alone. This is a result of the improved manufacturing and crystal technology used in current units. However, this fact is only valid if the transducer is used within its operational limits.

In general, drift effects can be minimised by the actual transducer design; therefore, selection of the transducer, its quality, and the suitability of the transducer design are imperative for a quality measurement. In addition, the actual transducer installation in the engine must be effected carefully, with consideration given to the mounting position and care taken when installing the adaptor. Note that the largest errors from the transducer alone are induced by engine related effects. These are

- Temperature—causing drift.
- Pressure and stress—causing deformation of the mounting bore and stress in the sensor housing.
- Vibration—causing noise on the pressure trace.
- Electrical interference from the engine and test-bed environment—causing electrical noise on the measured curve.

As mentioned previously, careful selection of the transducer for the application, in combination with appropriate consideration of the mounting position and careful implementation of the mounting bore, will ensure that errors induced by the transducer and its environment are reduced to an absolute minimum. This book contains a dedicated chapter on transducers, adaptors, and related issues and considerations for the combustion measurement application; please refer to it for more detailed information.

9.1.2.2 Cabling

The cabling between the transducer and the amplifier, in addition to general cabling considerations for piezoelectric measurement systems, is an often-overlooked issue. The cable requirements are quite specific and must be observed at all times with respect to the cable type between transducer and amplifier. In addition, the cable routing method is important to help with resistance to electrical noise, and cleanliness of the cable and connectors is imperative to prevent charge leakage. If possible, the measurement cables (carrying the charge signal) must be routed completely separately from other measurement system and power cables. They should be kept clean and dry and mounted in a fixed position so they are not subject to any vibration or movement, as this could generate electrical charges due to triboelectrical effects. If the cables or connectors are damaged, they must be replaced immediately to ensure reliability of the measurement system. Another crucial issue to consider is grounding effects. Differences in ground potentials between the engine, the amplifier rack, and the measurement device can cause circulating currents to flow along the cable shields. This will cause noise and interference on the signals acquired and can be very difficult to trace and eliminate. Equipotential bonding between all parts, via large cross-section grounding cables, is the most efficient way to ensure that all parts of the test cell ground are at the same potential.

Measurement cables can be repaired and their insulation properties restored by appropriate cleaning and/or drying. The shorter cables, normally attached to the transducer and supplied with it, are generally replaceable if found to be faulty. Connectors are generally consumables, and a small stock of spare connectors and adaptors should be held, to ensure no measurement downtime due to failure of these small, low-cost parts.

9.1.2.3 Encoder

Along with the transducers, the encoder is the other part of the measuring chain that is engine-mounted, and for that reason, it is a risk item that is likely to cause reliability problems and errors. The degree of this effect depends upon the encoder itself. It is always recommended that an encoder specifically designed for combustion engine applications should be used, whenever possible. Enclosed, optical types are generally most reliable, and these can be fitted to the front, non-drive end of the engine quite easily. Where this is not possible (e.g., high-performance engines), open encoder wheels can be incorporated, but these require more adaptation. In addition they are less reliable, as the encoder marks and optics are exposed to sources of contamination from the engine and test-cell environment.

An alternative is to use existing pulse wheels with sensors. These can also be susceptible to dirt, but more importantly they can suffer from phase shift, which can affect the raw data and thus the calculated values derived from the measured pressure curve. In addition, if a piggyback, parallel connection is made to an existing engine electronic sensor, care should be taken that noise or interference is not introduced onto this signal, as that could affect the engine electronic control or management system.

Most combustion measurement systems have a continuous monitoring of the encoder signal, counting the number of incoming marks and comparing with the number defined in the system setup, in real time. Errors are normally flagged in the user interface if they occur, so that they are immediately visible to the operator of the system.

Electrical noise on the encoder signal can cause errors and phasing issues. Most suitable encoders use fiber-optic signal transmission close to the engine and differential transmission for the electrical signal to the measurement system, and this provides good noise rejection in the hostile test-cell environment.

Engine torsional and resonance effects can affect the encoder signal. At certain engine speeds, the natural resonant frequency of the rotating engine parts can cause extremely high instantaneous speeds and rotational accelerations. This can cause the time difference between incoming marks to reduce to the point where the frequency of acquisition of the combustion measurement system is too slow, thus causing time-out errors that normally halt the measurement procedure.

The key to reliability for engine encoders is to use an appropriately designed unit and to ensure that it is properly installed, mechanically and electrically, into the test environment. Encoders should also be considered as consumable to some extent. They are often damaged, so a spare encoder or repair kit (including marker disk) is essential. This is particularly important where open-type encoders are used.

If sensors and trigger wheels are used, these can also be susceptible to contamination and may need frequent adjustment. However, they are normally relatively cheap, such that spare units can be easily maintained in stock.

9.1.2.4 Amplifier

The reliability of the charge amplifier, and its effect on the measurement quality, are largely functions of how well and accurately it has been set up. As it is not generally mounted at or near the engine, its operating environment is less harsh than that of the engine-mounted parts of the measurement chain. However, it is important to remember that the charge amplifier is an electronic instrument; it is therefore sensitive to electronic interference in a similar way to the pressure transducer and cabling. For this reason, similar precautions in the operating environment must be applied with respect to cleanliness of the connections, interfaces, and grounding, to prevent circulating ground loop currents. The effect of electrical noise on the measured data normally takes the form of transient spikes, of short duration.

In addition, the serviceability of the unit itself has an impact. Charge amplifiers generally work reliably and have a long service life. This means that in many installations, older amplifier units employing analogue electronic circuits are used, and these are much more susceptible to errors and inconsistent performance. This is due to age-related drift effects in the electronic circuit components themselves. Units of this type are even more sensitive to the external environmental conditions, and for that reason, they should be managed very carefully. This means that, in the operating environment, they must be kept away from sources of electrical interferences and maintained at a stable, consistent operating temperature. It is recommended that that rack into which the amplifier units may be installed should be left switched on, or at least allowed to condition for some time before measuring procedures start.

Newer amplifiers employ digital processor-based technology in the charge amplifiers. These are much more resistant to electrical interference and noise, as well as temperature-related effects. They are also resilient to electronic drifting, due to the fact that there are minimal analogue circuit components used in their construction. For these reasons, digital amplifiers can be mounted in the test cell, closer to the engine, thus reducing the length of the cables carrying the electrical charge. This is a clear benefit.

Apart from the operational environment, the other major factor that can cause errors in the measurements, which relates specifically to amplifiers, is the parameterisation or settings of the amplifier itself. Often, the settings are made via switches or dials on the front panel of the amplifier, and there is always a possibility that these may be wrong, or that previous settings used may be untraceable. The only solution is to instil a strict regime of traceable records for the system setup for every measurement task. This will ensure that any errors are likely to be corrected at the source—or, if not, at least traceable for the possibility of correcting the data in postprocessing.

9.1.2.5 Measurement Device

Analogous to the amplifier, the measurement device generally operates in an environment where it is relatively well protected. That is, it is normally in a control room, with stable temperatures and minimal electrical interference fields. The main areas for potential problems relate to the parameterisation and optimisation of the system itself. When the system is configured for use, attention must be paid to ensure that the best compromise of the system's limitations or boundaries is applied for the application. Also, many systems incorporate features to optimise and improve the efficiency of the measurement processes and to reduce the requirement for handling the large amounts of data that combustion measurement systems often generate. Users of the system should familiarise themselves with these features, in order to get the best system setup for the task to be undertaken. System parameters must be carefully stored, managed, and controlled, to ensure that the correct parameters are loaded for the specific engine and test. A structured, traceable approach should be applied to the measurement process, to ensure that a procedure for pretest checks and test-data validation is applied and is effective.

Often it is useful, and good practice, to take reference measurements at the beginning of and during a test program. This allows a validation of the data quality without disturbance of the instrumentation, to ensure that the measurement chain is performing accurately and repeatably. A typical process would involve engine conditioning (warm-up procedure), taking measurements at reference operating points, checking that the boundary conditions are fixed to a reference value (fuel type, oil and coolant temperatures, etc.), checking that the measurement system setup meets the reference conditions (parameters and settings), then executing reference measurements, paying particular attention to sensitive measured values and parameters such as IMEP and peak pressures. Once several sets of measurements have been collected over time, tolerance limits can be set, based on the chronological data. This reference measurement can be taken on a regular basis to prove measurement chain and data quality. Any trends over time will be visible and can be used as a measurement of the life of consumable components (transducers) or as an indicator of problems or errors in the measurement chain

9.1.2.6 Summary

The main components in the measurement chain, with respect to the instrumentation error effect, are classified in the table shown in Table 9.1.

Table 9.2 classifies measurement chain components with respect to their sensitivity to external interference.

Table 9.1 Instrumentation Error Effect

	Transducer and cabling	Amplifier	Encoder	Measurement device
Drift	Significant	Significant	None	None
Linearity	Significant	Some	None	Some
Stability	Significant	Significant	Some	Some

9.2 Successful Setup and Diagnostics

9.2.1 Basic Setup

The most important point to remember when setting up the system is that the acquisition process is relatively simple. Whatever system is used, its main function is to measure and store in a digital form, at an appropriately high resolution, the measured curves from the instrumented engine. These are normally cylinder pressure curves, but may also include other high-speed, angle-based channels that relate the combustion process. Normally the digital samples are triggered via angle encoder pulses; hence, the data will be in the angle domain.

There are many software and hardware features in a modern combustion measurement system that have been designed to support the application. Also, it may be possible that a simple, generic data-acquisition card may be used to measure and store the data, but the key to a good-quality measurement is to ensure that the system is setup is correct and appropriate, taking care to ensure that full optimisation of the system features has been applied. This will ensure the collection of good-quality, digitised data.

There may be many convenience features in a specific combustion measurement device, but the most important factors to consider when collecting or analysing the data are relatively simple:

1. Correct pressure referencing of the measured curve on a cycle-by-cycle basis. This is often known as "pegging" or "zero-level correction." There are methods discussed in this book that can be applied in real time, during the acquisition process. Alternatively, the zero level of the measured curve can be adjusted in postprocessing from measured inlet-pressure data. However, it is essential to directly reference the pressure curve,

Table 9.2 Sensitivity to External Interference

	Transducer and cabling	Amplifier	Encoder	Measurement device
Electrical noise	Significant	Significant	Significant	Significant
Vibration	Significant	None	None	Significant
Temperature	Significant	Some	Some	Some

both for direct calculations but, more importantly, also for the indirect calculations derived from the pressure curve, particularly the energy release and polytropics.

2. The raw data curve, as measured, has to be correctly calibrated and offset in the Y direction (i.e., pressure referencing, described above), but just as importantly, in the X direction—more commonly known as TDC referencing or phasing. It is most important that the TDC position of the engine is correctly established and applied to the data either during or after the measurement. TDC determination methods have been discussed elsewhere in this book. This factor is critical with respect to direct and indirect calculations, in particular, those calculations that require accurate cylinder volume assignment, for example IMEP.

3. Engine parameters are needed for accurate calculation of the cylinder volume with respect to crank angle. Generally these will be known or available; the main source of error relates to the difficulty in accurately establishing the clearance volume/compression ratio, particularly with a prototype engine or a high-compression engine, where manufacturing tolerances may become significant. Also, deformation, due to thermal and mechanical effects when the engine is running, may be difficult to quantify accurately.

4. The engine compression and expansion process definition (polytropic index) is used for calculations relating to online pressure referencing. In addition, it is used in the standard fast energy-release calculation in most combustion measurement systems. For this reason, it must be accurately established. This can be difficult, as the value will change throughout the cycle, depending upon the engine operating conditions. As a general rule, the value should be defined in accordance with which part of the measured data or application is the most important with respect to accuracy.

9.2.2 System Prechecks

9.2.2.1 Introduction

A methodical approach, and some time spent on prechecking the measurement setup, will reduce the likelihood of errors in the data that could have easily been avoided. This improves the measurement process efficiency and reduces the probability of test down time due to combustion measurement system issues. A typical check procedure is suggested below, working backward from the engine to the data acquisition system. A disciplined prechecking procedure should be applied at the start of each measurement application. In addition, a reduced quick check procedure can also be applied that can be carried out more frequently, to capture any potential problems before they halt the measurement.

Transducer and Cabling

- Make sure that the transducer has been correctly mounted in the adaptor and the correct torque figure has been applied during installation.
- Double-check that the correct type of transducer has been selected for the measurement application.
- Ensure that calibration and operation information is available for the transducer. Its last calibration date and calibration data should be available, also. If possible, confirm

the transducer operating life and conditions thus far: it would not be prudent to start a test procedure with a unit that may be near the near the end of its life. Alternatively, if it has been used previously in harsh measurement conditions (e.g., knock measurement), it may no longer be suitable for accurate measurements.

- Check that the transducer cabling around the transducer and engine, then to the amplifier, has been appropriately laid. Cables must be off the ground and away from the risk of contamination that will cause charge leakage and signal drift. Any damaged cables or connectors must be replaced.

- If a cooling system is needed, make sure that the pipework is correctly routed. Check the sight glasses while the system is running, to ensure flow to each transducer. Make sure that the fluid level is at the correct operating level, and observe the system for leaks during engine run time.

- Check regularly for pressure leakage at the transducer. If possible, use copper sealing gaskets, as these are more reliable.

Amplifiers

- Ensure that the amplifier rack is appropriately located, according to its operating specification. This is particularly important with respect to temperature and noise immunity.

- If possible, leave the rack switched on, or allow it to warm up, condition, and stabilise before starting measurements.

- Check that the amplifier settings are correctly assigned at the amplifier itself. This is particularly important for amplifiers that do not have an intelligent interface to the operating system computer for the combustion measurement system. It is always possible that a switch could inadvertently be set in the wrong position, and this may change a parameter that may not be that visible during the data acquisition process. Pay particular attention to the gain and drift compensation settings. These will have the greatest impact and are most likely to be assigned incorrectly, as they are generally adjusted from the front panel of the amplifier.

Encoder

- Inspect the angle encoder mechanical installation. For enclosed-type encoders, normally mounted at the free end of the crankshaft, make sure that the unit is securely mounted, via the supporting or mounting arrangement, to the engine itself (not to the mounting frame that the engine sits on). Check that the installation is as vibration-free as possible. If necessary, check the encoder runout (eccentricity of mounting) via a dial gauge, as any radial eccentricity produces an additional angle error during measurement. For open encoders, normally mounted at the drive end of the crankshaft, a bespoke manufactured disc will be produced, according to the mounting position. Make sure it is correctly and securely clamped according to fitting guidelines. The disc must be accurately centered, with minimal runout. The optical pickup must be mounted to the engine block as rigidly as possible, to prevent measurement error resulting from relative movement between the marker disk and the pickup.

- Inspect the electrical connections and wiring; make sure these run away from the engine's main heat sources (e.g., exhaust system) and, if possible, away from sources of electrical noise (e.g., ignition system). If possible, run cabling off the floor. Any intermediate components—for example, pulse multiplier boxes or converters—should be securely mounted, off the floor on the engine mounting frame, where they are accessible for diagnostic purposes.
- Check the setting of intermediate hardware units between the angle encoder and measurement device. Make sure that these are configured appropriately to produce the required mark resolution as expected by the measurement device. Certain combustion measurement systems normally have a diagnostic and measurement procedure to support correct configuration of the encoder pulses. Use this to check for the correct incoming crank degree and trigger marks before the actual measuring procedure.

Measurement System

- Make sure that the correct parameter file settings are loaded for the test. Use the system software features to check correct function of the incoming signals.
- Make sure that the analogue input range of the system is fully utilised, to ensure good-quality digital conversion. The maximum expected signal input value is needed to establish this. In addition, check that the charge amplifier or signal conditioner settings are appropriately configured to support this (for example, by applying a zero-level offset). Make sure that there is no possibility of overshoot (overvoltage applied to the input channel), as this will halt a measurement procedure as a system error.
- Check that TDC establishment has been carried out and stored in the parameter settings to be used. If the angle encoder has been removed or is newly installed relative to the test environment, this TDC measurement process will need executing.
- Check that all cabling and cable interfaces to the measurement system are secure and correctly located.
- Check that the number of cycles required to be measured and stored is possible, based on the memory size and number of data points per cycle. Check that the maximum required engine speed can be achieved, based on the data throughput rate, which is a function of the required measurement resolution and digital conversion speed of the system analogue-to-digital converter. In addition, the channel count has an effect if the analogue to digital converter is multiplexed across channels.
- Do a quick quality check by measuring a few motored pressure curves. Examine the curves closely: poor-quality digital conversion will be seen as a curve with steps (motored curves are low-amplitude and therefore worst-case). Check the position of the curve peak, which should be at TDC if the definition of TDC is correct. This can be checked even more closely via a log PV diagram (discussed in the next section), for absolute accuracy. Examine the pressure value around inlet valve closure, which should be around 1 bar absolute. If not, check and/or apply the zero-level correction method. This measurement of a motored curve provides a quick "sanity" check, as it highlights many simple but often overlooked issues. Save this measurement data as a reference point for future diagnostics if needed.

9.2.3 Diagnostic and Reference Measurements

Time spent measuring and then analysing acquired raw data curves during the system setup process can provide a good understanding of the quality of that particular setup. Saving some data that is known to be of good quality, prior to the start of a test program, allows the establishment of a baseline that is traceable, and this is useful for diagnostic purposes. In addition, it is a good confidence indicator to the data analysis engineer or scientist that the measured data is of good quality and has been collected by a individual who is capable of clearly identifying any potential problems. Test environments vary. Some users of a combustion measurement system will collect data for an engineer, who will then examine that data at a later time. In other environments, engineers themselves will collect the data during test, then immediately analyse it. Irrespective of who will be processing the data in the end, some quality checks during the measurement process and some good-quality data, measured and saved at the beginning of the test, can provide a useful metric for comparison use as the test program progresses. The question is, what measurements should be made, and what are the important observations to be made on this data? We have mentioned before the main areas of concern: correct TDC allocation and pressure referencing (pegging), to ensure that the pressure and volume relationship of the measured data is accurate. In addition, critical thermodynamic and engine parameters must be correctly defined, and the acquired data channels optimised for good-quality digitisation. Therefore, the diagnostic measurements should reflect and identify any issues or problems with these main areas of concern. The fundamental thermodynamic relationships between pressure and volume will always hold true. Measurement of motored and fired pressure curves, which are then examined as a pressure-volume diagram with log scales, is a common approach to verify the quality of the equipment setup. A diagram of the pressure data will highlight errors in many quality-related criteria. A typical logPV diagram is shown in Figure 9.1.

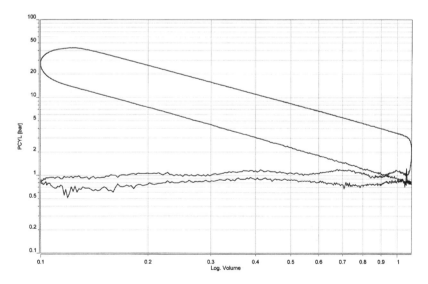

Figure 9.1 Log PV diagram for a fired engine.

(Source: AVL.)

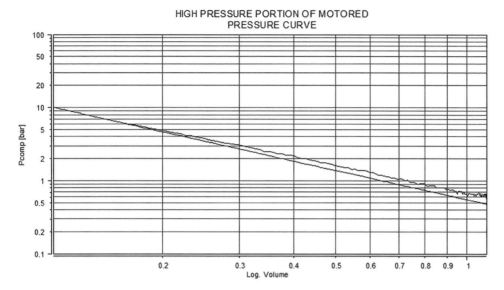

Figure 9.2 High pressure portion of LogPV diagram for a motored engine.

Analysis of the motored logPV diagram is an important task for quality checks. Motored data is very repeatable and does not suffer cyclic variation, unlike fired data. The compression and expansion polytropic processes can be derived directly from this diagram, as they are represented by the gradient of the slope. In theory, this slope should be a completely straight line during the process, but deviations from the straight line can be useful diagnostic indicators of issues with the measurement chain.

In theory, an ideal engine would have a motored logPV diagram in which the compression and expansion lines are completely overlaid, with no enclosed area—and hence no work done or lost. Of course, in the real world, this is not achievable—but interestingly, the errors in the pressure-volume relationship, caused by typical errors in the measuring chain that we have mentioned, will be clearly visible on the motored logPV diagram.

Incorrect TDC assignment will clearly affect the pressure relative to volume. Errors in this respect can be clearly seen on the motored logPV plot as a curve inflexion or crossover just prior to the peak at TDC (Figure 9.2 and 9.3). This characteristic shape/inflexion of the curve is only possible due to an error in TDC definition and is hence a clear and reliable indicator of this problem.

Incorrect pressure referencing of the pressure signal will cause curvature on the compression and expansion curves, although this is more pronounced during the earlier part of the stroke. An upward curve (end curving up) indicates that pressure of the referencing value is too high; the opposite for a downward curve. The effect on fired curves is shown in Figure 9.4.

Incorrect definition of the compression ratio/clearance volume has a similar effect on the pressure-volume relationship, causing a distortion that affects the higher-pressure part of the cycle around TDC (assuming that the reference pressure is correct).

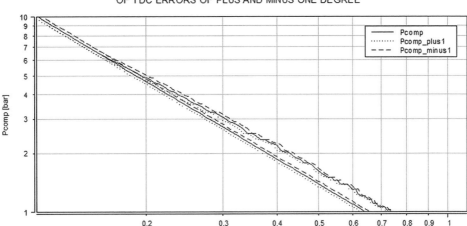

Figure 9.3 Peak of motored logPV diagram showing the effect errors of plus and minus 1-degree crank angle.

As mentioned, the gradient of the high-pressure curves is the polytropic index of the compression and expansion process. These, of course, have some variation due to specific operating conditions, but they are known to fall within a realistic tolerance band (1.25 to 1.37). If they lie outside of this range, then scaling of the pressure data is an issue. In this case, the transducer calibration and signal-conditioning settings should be checked.

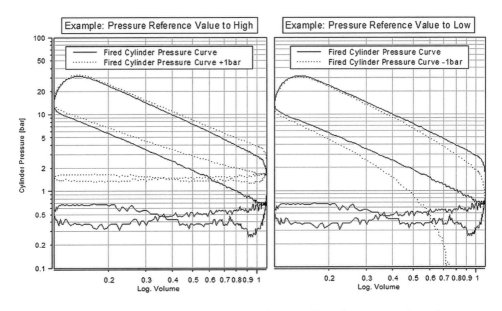

Figure 9.4 Fired log PV diagrams showing the effect of pressure-referencing errors.

Assuming that the motored data has correct TDC and scaling, irregular curvature or sloping of the compression and expansion lines could also indicate transducer problems.

Motored pressure analysis is a useful approach, but ultimately the measuring chain is still not operating in a realistic condition that would highlight all potential problems. When motoring, the transducer temperature is significantly different than in fired conditions, and thus temperature-related effects (like thermoshock, short-term drift) will not be apparent. It is therefore necessary to examine other metrics, under fired operation, that could help highlight potential problems.

There are a number of simple checks that can be made on the fired data as indicators of correct setup, particularly with respect to the main problem areas that we have discussed.

A simple examination of the pressure curve itself should show clearly any electrical noise present. The curve should be smooth, and at low amplitude should not display any "stepping" that could indicate low-quality digitisation. For correct pressure referencing, look at the curve values around inlet valve closure. If manifold pressure values are measured or can be estimated, compare them with the actual value. Any significant deviation could highlight a pressure referencing error. Make sure this is observed at low speed and load, as gas dynamics at higher speeds will falsify this evaluation.

If these checks prove acceptable, then examine some calculated results, in particular those that rely on an accurate pressure and volume relationship. One can always assume that IMEP should always be greater than BMEP. In addition, IMEP for the high-pressure part of the cycle should be greater than IMEP for the gas exchange part. It seems obvious, but simple checks like this are often overlooked. From the pressure and volume, the fast heat release integral curve can be derived. With many combustion measurement systems, this can be displayed during measurement, and the general shape of this curve gives a good indicator of measurement data quality. The curve should conform to the generic S shape. In particular, it should have a flat plateau toward the end of energy conversion; any slope here could indicate errors in system configuration or parameters. If this is not the case, in conjunction with other quality indicators that we have mentioned, the measurement system needs attention before embarking on the measurement program.

Assuming that the system checks and reference measurements show that it is performing correctly and produces good-quality data, it is worthwhile to measure and store both fired and motored curves, as reference data for ongoing quality checks that should be made periodically throughout the test program. The results of these should be summarised and monitored on a tracking sheet so that deviations or trends in the reference measurement data quality, which could indicate failure or incorrect setup, can be identified quickly.

9.3 Software and Data Handling

9.3.1 Introduction

The most important output from the measurement system is data—as in most cases that is the only output. For this reason it is important that the data produced is accurate, reliable, stable, and repeatable for a given set of operating or measuring conditions. In addition, the software used in the measuring system has a life cycle. In order to ensure the reliability of the hardware, the control and user interface software for the combustion measurement systems must be maintained, updated, and controlled. In addition, any personal computers that are part of the system must also be managed appropriately and updated when necessary. For these reasons, software and data management is an important topic to ensure the reliability and usability of the measurement system. In addition, good practices with respect to measured data and parameter settings management will ensure that the working environment is productive and that all activities are traceable for the purpose of data quality and integrity.

9.3.2 Measured Data Requirements

The main issues to consider with respect to the data produced by the combustion measurement system relate to the nature of the measurement task itself. These issues are as follows.

- The system throughput rate is high. The amount of data to be stored from a typical measurement, over tens or hundreds of cycles, produces large data files that require significant mass storage capability.

- The data produced needs a complex data model to support all the required information—for example, multiple channels with different measurement resolutions, calculated results for these channels, statistics of those results, volume table information, engine parameters for generating the volume table, etc. This requires a complex format for which, at the time of writing, no standardised data model exists; hence, most manufacturers have their own binary file format, all of which vary greatly.

- Data files and parameters need a logical connection or link; otherwise, it would not be possible to know some of the important parameters needed for data evaluation or correction—for example, zero-level correction method used, TDC offset, and calibration settings.

- Often, measurements from the combustion measurement system need to be correlated with engine test bed measured data. This is particularly important where the test bed system has a remote connection and/or measurement trigger for the combustion measurements. A strict regime or routine is required with respect to file-naming convention, or other methodology, in order to be able to easily identify and correlate measurement data from the different systems. Some remote control connection protocols for combustion measurement systems have auto–file naming and parameter transfer between systems to support this requirement.

These factors mean that combustion data is often difficult to handle due to large data files, and difficult to evaluate for the same reason. Most suppliers of data acquisitions systems will provide or recommend software tools for viewing and printing the data. In addition, most allow export of data in alternative formats for viewing in commercially available analysis and visualisation programs. However, many scientists and engineers are very familiar with office-based programs for calculation and data visualisation. The question is, what is the best approach for the following general tasks:

- How to store the data, where, and what format?
- How to visualise it quickly after the measurement?

- How execute further calculations for verification and advanced analysis?
- How to correlate and synchronise measurements with other systems, or how to know what measurement was made when?

As manufacturer equipment varies considerably with respect to format of the data, we can only discuss some generic guidelines for these points. However, this is the best approach to take as, often, a test field may contain equipment from various suppliers, or even in-house equipment. Therefore, the following general rules could be applied, in the interests of achieving best practice:

- For commercially available equipment, it is recommended to store that data in the supplier's native format. Often this form contains all the information needed for data visualisation and further analysis. This format is normally a binary format and should be checked for its storage efficiency. If storage space is an issue, compress the files with common file compression tools; the resulting compression ratio also gives an indication of the file storage efficiency.
- Check that the file contains some logical link to the parameter settings that were used. If it is particularly important that calculation of the volume table is possible in postprocessing, this requires a channel containing crank position, as well as the appropriate engine geometric parameters.
- When storing files used in a particular measurement series, use an appropriate file structure, containing main folder with subfolders for data, parameters, calculations, scripts, notes, etc. This will assist in understanding the data and the measurement when analysis takes place at a much later time.
- For quick visualisation of data, use the native file format in conjunction with the supplied or recommended software tool from the combustion measurement system supplier. This task is often executed immediately after the measurement, for plausibility and integrity checking.
- If data is required for export, consider what it is needed for. If it's just for a simple picture in an office document, use a screenshot export. If further analysis is needed, use a standard, delimited format that can be read easily by other programs. Note, though, that cyclic data in this form creates very large files; when using this format for raw data curves, it is highly recommended that averaged curves or curve envelopes only are exported. For calculated results, these are normally tabular in form, so they are ideally suited for viewing in spreadsheet-type programs. Remember that most export formats do not support the inclusion of parameter data, due to the complex data model associated with combustion measurements. If complex analysis is really needed in an external program, consider writing a script or a program DLL (dynamic link library), to be able to read the native data format in the chosen software environment. Most equipment suppliers will provide the data file structure so that a software interface can be created. Although this requires some initial investment, this can be justified where a lot of postprocessing and analysis is anticipated to be done via alternative calculation and visualisation programs.
- Where measurements have to be correlated and the data aligned between the combustion measurement system and other measurement devices, consider what

information is to be exchanged between systems. How will the files be identified, with respect to the measurement condition? Will they be steady state measurements or transient? An appropriate extended file name convention is useful, as it allows the measurement condition information to be visible without having to actually open the file itself.

Efficient handling of the data files themselves will promote a productive environment, but this depends on the actual requirements of the person who needs the data from the combustion measurement system.

A typical scenario for a small testing environment would be that all measured data is saved locally on the combustion measurement instrument computer. File names and measurement conditions are manually noted. Once the test series is complete, the data files are transferred to an office computer environment for processing and evaluation. This methodology is suitable where combustion measurement is a detailed investigation, not being carried out as a routine measurement. Often the engineer or scientist who needs the data might supervise or even execute the measurements, then copy the data themselves to their own environment for processing. A typical example of this would be single-cylinder research engine-based work, fundamental investigation of engine combustion and related processes. In this environment the combustion measurements, in addition to other advanced measurement techniques, are the main focus.

For a larger test field, where multiple test beds exist, the requirement needs better management. Often, in this case, multiple measurement systems coexist; it is important that data management is efficient, to facilitate easy access to the data. In addition, all measurements must be traceable to the source measurement device. In this environment, where test beds often run automatically, the combustion measurement system is an integral part of the cell infrastructure. Often each cell has its own combustion analyser. An intelligent approach here is to use the test bed automation controller to trigger the combustion measurement system operation. In this way, measurements are executed in conjunction with all the other measurement devices, and the related combustion data is transferred to the test bed host data environment. This is very typical for calibration test environments where the combustion measurement system is essential for the task. A typical process would be as follows:

- Measurements of multiple cycles of raw data and results are executed automatically, controlled by the automation system. Raw data and results are to be saved locally and backed up regularly to a host file server.
- Statistics of the results are calculated after the measurement and transferred to the test bed data model for storage with other test bed data channels.
- File naming convention is controlled by the test bed automation host, each log point having a file name for the combustion measurement data. In this way, the raw data from each log point is traceable during processing.

Another possibility is that the combustion measurement system has an online, real-time link to the test bed automation or host system for transfer of results during measurement. The system continuously measures data, calculates results, and transfers them to the host (i.e., test bed automation system), actually behaving in an unintelligent way—simply transferring data. The host system considers this online data as a simple measured channel (or number of channels) that are continuously sampled and stored with the other measured channels, in either transient or steady-state test mode. The interface connection does not generally allow remote control of the combustion measurement system; the combustion measurement system simply operates in the background, continuously polling new data values onto the interface. This interface could be analogue, via digital-to-analogue converter channels, or it could be a digital interface—for example, via TCP/IP or CAN. The advantage of this technological approach is that it is simple to set up at the combustion measurement system, as well as the host side. The data channels, of continuously updated cyclic results, are measured and stored easily as test bed channels in the automation system. The disadvantage is that raw data is not generally saved with this technique, so any errors in results cannot be investigated further by examining the raw data curves from which they were derived. Also, depending on the interface performance, there may be a restriction on bandwidth if transferring a large number of result values (e.g., 8 cylinders, 6 results per cylinder = 48 results per cycle). Or there may be a restriction on the number of output channels available. (Digital-to-analogue converters are often limited, at 16 outputs for combustion measurement systems.)

9.3.3 Plausibility Checks and Data Validation

Previously in this book, we have mentioned the topic of data validation and reference measurements. This is a very worthwhile activity to monitor the quality of data during a test program. There are two main areas to consider in this respect. The first is validation of the measurement immediately before, during, or after the measurement. This involves the techniques that we have already mentioned in this chapter, closely examining motored and fired data curves and the results derived from these, checking them for plausibility, and using experience and knowledge of the measurement task to monitor correct parameterisation of the system and to ensure that external interference effects can be reduced or removed so that they do not corrupt the data or measurement task.

The second area to consider is the integrity of the data over the test program life cycle. Often, an engine will be installed on a test bed for a period of time while development or calibration tasks are executed. In this case, the repeatability of the combustion measurement system is an important factor (as, indeed, for all the other measurement systems in the environment). For this reason, reference measurements—made regularly, and routinely plotted and monitored—become a significantly helpful tool to understand the source of any variation in the measured data. That is, is the variation natural, or has something changed to cause it?

Reference measurements involve measuring some combustion data, at set or known conditions, then reducing the data to a simple indicator value that can be plotted regularly on a monitoring chart (i.e., average or statistic of a result value). This should be done regularly (daily or weekly) at an engine condition that is representative of the conditions that are of most interest during the actual test. An example, perhaps, would be to measure motored data at normal engine temperature with WOT, and then fired data at a part-load, medium-speed condition. Every time this reference test is executed, the boundary conditions must be exactly the same (as much as possible). For example, the same controlled temperature conditions (oil, water, etc.). Perform the same warmup procedure; use the same order for executing the reference measurements; if possible, have the same person execute the test. This ensures that any small variation in the data is highly likely to be natural. Large variation highlights an error that needs investigation or understanding.

Once a statistically representative set of measurements has been made, critical, sensitive values can be plotted on a regular basis, to begin to understand the variations in this particular test setup. The typical elements of interest would be MEP values from the combustion measurement system, as well as peak pressure-related results and load values from the test bed (torque, manifold pressure, BMEP).

Once an appropriate number of data samples has been gathered (5 to 10, or more), natural variation will be visible, and controlling limits can be defined and set based on this data. An appropriate technique is the control chart (developed by Walter Shewart): the upper and lower control limits are defined as a function of the standard deviation of the data set, using plus or minus 3 standard deviations to set the limit values. Note that it is crucial that the initial reference data set contains only naturally occurring variations. An example chart is shown in Figure 9.5.

This chart plots the engine average IMEP at the reference condition on a daily basis. Any variations can be clearly seen (Figure 9.6), and this can facilitate immediate investigation of the system to ensure a clear understanding of the phenomenon. It could be due to an error condition or a faulty component (e.g., a failing transducer) that needs replacement. Note that if a component needs replacing or recalibrating, on either the engine or the test system, then the control limits will need recalculation. The control chart will also highlight trends that may stimulate investigation procedures, even before control limits are exceeded.

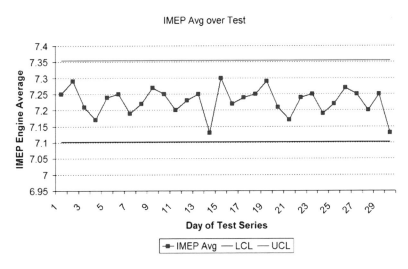

Figure 9.5 A control chart for engine average IMEP data over a test.

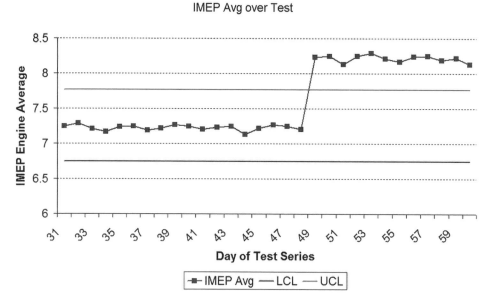

Figure 9.6 A control chart showing an unexplained deviation that would indicate that a deviation has occurred and needs investigation.

9.3.4 Best Practice and Summary

It is important to consider that the combustion measurement system, as well as the complete test environment, is an expensive tool for engine and powertrain development. The equipment is specialised and the test bed demands large quantities of energy and manpower. The only thing that a combustion measurement system produces (as is the same for a complete test facility) is data. Therefore the data produced is an expensive commodity that must be of very high quality and must be reliable and repeatable.

The data produced by a combustion measurement device is essential in the engine development process. The data sets have a complex format, due to the nature of the measurement, and are generally large. This means that accessing the data, managing it for processing and display, and archiving it can be a considerable task that needs a robust data-storage concept and methodology, particularly in a large test field or facility. Another complication is that data can be collected, stored, and exported in various formats, and in addition, there is no standard format for combustion measurement data. This adds further complexity to the challenge of storing information and distributing it to the people who require it. It is a challenge, but not insurmountable; many manufacturers of combustion measurement equipment have tools and technology to assist in this task. The main consideration is the existing environment and processes, examining where improvement could be made, and then defining the pathway to achieve the improvements in an interactive way without causing confusion or conflict.

Process improvements for gaining higher-quality data from better-controlled procedures is a very worthwhile objective. For combustion measurement applications, reference measurements and associated control charts are a powerful tool, but the underlying processes must be executed and strictly enforced. The reference measurements must be executed reliably, and the data must be plotted for review regularly. Prompt action should be taken when an unexplained deviation occurs, but control limits must only be reset when a change in the test environment occurs (for example, fitting for a new transducer, change of engine component). In addition, it is important to never make a decision based on a single event or change; make sure that you fully understand the natural variation present in the combustion measurement chain and test environment.

9.4 Hardware Handling and Maintenance

9.4.1 Introduction

The combustion measurement system hardware consists of the data acquisition system and computer, the amplifiers, plus the cabling, transducers, and encoder. We have already discussed the possible errors and problems that can be associated with hardware-related parts of the measurement chain. In addition, we have mentioned best practice with respect to location and layout of transducers and cabling. In this section, the main focus is repair and maintenance of the complete system, most particularly the engine-mounted parts, as these are subjected to the most extreme operating conditions with respect to temperatures and dynamic forces.

9.4.2 Measurement Hardware and System

The measurement system hardware consists of the data acquisition system, with associated user interface (normally a personal computer), in addition to the amplifier hardware, which normally consists of individual amplifier modules mounted in a rack or cabinet.

The data acquisition system generally needs very little hardware-related attention during its operating life; modern measurement system hardware is quite reliable and only gives problems if it has been operated outside its specified working environment tolerances—for example, at low or high temperatures or humidity levels. This is not generally the case, though, as the measurement system is most often located outside of the working environment, at the test bed control interface, not inside the test cell itself. The routine maintenance generally required for a system involves the following:

- Calibration check of the input and output channels.
- Check of correct operation of any cabinet ventilation or cooling system to maintain correct operating temperature.
- Check of software version; installation of hot fixes or updates.

Calibration and linearity checks can be executed by the system user/owner via the use of high-quality, traceable voltage sources and measuring instruments. If a channel is found to be faulty, the equipment generally must be returned to the manufacturer for repair or replacement. This is also the case for software maintenance; the manufacturer should supply software and firmware upgrades. These should be installed, and then the system checked for correct function and operation. It is important that a system user should be aware of any available upgrades or hot fixes so that these can be installed as soon as possible. Manufacturer websites generally provide this information.

Where systems are used for in-vehicle measurements, often, this involves measurements at temperature extremes. In addition, the measurement system, mounted in the vehicle itself, is a considerably less stable environment. The measurement device is more likely to be exposed to vibration and shock. Care should be taken, when mounting the system, that it is secure and not likely to move around during the test procedures. In addition, consideration should be given to the operating environment. For example, in extreme cold, the unit should not be switched off; leaving it on will prevent condensation, which would destroy electronic components, from forming inside the equipment.

Amplifier racks must also be maintained to an extent. If the amplifiers are mounted collectively in a rack, the cooling fans and filters that are often fitted must be checked for correct function, to ensure that heat does not build up that could affect the operation of the amplifiers. We have already mentioned the necessity for cleanliness of cabling and connection interfaces; this is an ongoing requirement—continual maintenance to support quality measurements. Amplifiers often need calibrating, and this should be done on a regular basis—in particular, before embarking on an important test program. Analogue amplifiers tend to suffer from drift as they age; they can be checked as part of the calibration process for the whole measurement chain. Alternatively, they can be calibrated stand-alone. If a problem occurs, they would normally be either returned to the manufacturer or scrapped, depending upon their ages. (Amplifiers and racks often have long operating lives: 20 years or more.)

Digital amplifier racks are less sensitive to temperature extremes and environmental conditions. They are less likely to fail or drift, but nevertheless it is important that they are calibrated regularly as part of the measurement chain. The same issues apply with respect to rack cooling and cleanliness of the interfaces. An additional requirement for digital racks, similar to that of the measurement system, is that they may need periodic software or firmware upgrades, the source for these being the manufacturer.

9.4.3 Engine-Mounted Equipment: Encoder

Angle encoders are quite vulnerable, due to the mounting location at the engine. They are exposed to extreme heat and vibration, and therefore are a part of the measurement chain that is at risk of failure. They are many types of encoder solution that could be used; encoders are generally a low-cost purchase, and hence, failure of these units would necessitate replacement rather than repair. The same applies to trigger wheels and sensors; the actual sensor is low-cost and if suspected as faulty, would be replaced. Due to low cost, with these solutions, it is recommended that spare units be kept in stock at all times, so that failure does not cause extended test cell downtime. The main issue to consider is that low-cost units are less accurate; in addition, if they are not durable, they will be expensive with respect to the downtime that they may cause due to repeated failure and short operating life.

For greatest accuracy, there is no substitute for an angle encoder specifically designed for mounting on a combustion engine. These are available from several manufacturers of combustion measurement equipment and generally provide the best combination of value and performance in the application. In most cases, an angle encoder of this type will be repairable and serviceable. It is very rare that a unit of this type fails beyond economical repair, and often if it does, the failure is due to misuse or abuse. The main issue to consider with these units is correct installation. The mounting of the unit is crucially important for its life and accuracy, and therefore it is best always to follow the procedures specified by the manufacturer for mounting the unit on the engine and checking for correct alignment. In many cases, these units are trouble-free; they work until they fail, and then they are investigated. There is no specific preventative maintenance other than regularly checking the soundness of the installation (mounting bolt torque, visual inspection, etc).

Most combustion measurement angle encoders use optical signal detection of the marker disc. This can be an area that causes problems due to the environment in which they operate. Often dust and fluids can seep past sealing arrangements and contaminate the marker disk and optical sensors. This causes synchronisation errors or failure of the measurement system; regular cleaning of the marker disc and optics is therefore often needed. This is particularly the case with open-type encoders fitted at the flywheel (or drive end) of the engine, as they are completely exposed to their surroundings. Cleaning of the disc should be executed carefully, with no use of harsh solvents. Lint-free cloths should be used, to avoid fiber contamination. If disassembly of the encoder is necessary, refer carefully to the manufacturer's instructions. Certain parts of the encoder should not be dismantled, due to optical alignment requirements. In addition, correct lubrication products should be used upon reassembly. This is particularly important if the encoder is to be used for cold-start testing environments.

A procedure of regular checks should be carried out to ensure long life of the unit, but more importantly, to help ensure accurate measurements (see Table 9.3).

In addition to the regular checks, to ensure maximum uptime, a suitable spare-part contingency should be considered. This is particularly important where marker discs have a bespoke configuration and are made to suit specific requirements or dimensions.

Table 9.3 Suggested Preventative Maintenance for Combustion Measurement Angle Encoders

Check	Frequency
Function check of encoder, via measurement system, for correct phasing and synchronisation	DAILY, or before each measurement series
Visual check of encoder installation and cabling layout, security, and contamination	DAILY
Check of encoder installation for security of mounting and vibration resistance	WEEKLY, and before each test or measurement program
Dismantle of encoder, internal cleaning of marker disk and optics, re-grease	MONTHLY, or before each measurement program, or as required during a test program

A suggested spare-part stock for encoders, in an environment where more than one of the same encoder type exists, is as follows:

- Marker discs (closed type)—one spare to for each type/unit.
- Marker discs (open type)—one spare of each design (normally one specific design per engine)
- Repair kit (closed-type encoder)—one
- Pulse multiplier—one
- Specific connector/interface plugs—one spare of each for making interface/patch cables

9.4.4 Engine-Mounted Equipment: Transducers

9.4.4.1 Introduction

Piezoelectric transducers should be considered as precision measurement devices, and should always be treated as such. They are expensive consumable items that have a decisive effect on the overall performance and quality of the measurement system. In order to get the best performance and accuracy, they should be handled very carefully during usage, ensuring that they are always operated within their specified limits and tolerances, and protected from mechanical damage. This section describes the main issues to consider in the daily use and operation of piezoelectric transducers for combustion measurement.

9.4.4.2 Installation and Handling

Assuming that the sensor installation site or method has been chosen, there are certain accessories available that can help with installation of the adaptor and transducer. An important point is that where tools are available for a specific purpose—for example, installing the transducer in the adaptor, specific drill bits for drilling the engine for access—these should always be used. Most manufacturers of transducers provide a selection of the appropriate tools and equipment for the purpose.

Once the installation bore or adaptor has been chosen, it is most important to ensure that the correct mounting torques, on the transducer and adaptor, are applied. This is of critical importance, as incorrect installation in this respect can destroy the transducer, but still more important (and less obvious) is that fact that incorrect torque will alter the preloading and stresses in the transducer housing assembly. This will cause an unpredictable change in the sensitivity of the unit, which would directly affect the accuracy of the measured data.

When the transducer is removed from the engine, it should be handled carefully and stored appropriately in its original case (as supplied by the manufacturer). This will help to reduce the possibility of mechanical damage or shock. Note that it is particularly important to protect the diaphragm and sealing faces of the transducer during handling.

Most transducer designs are available as dummies, and these should always be used when the engine needs to be operated without the need for measurement (by replacing the actual transducer itself). During removal, transducers can become "stuck," due to deposit buildup.

This can have disastrous consequences for the transducer and bore, and therefore steps should be taken to prevent it. Typically this would involve:

- Cleaning the transducer and installation bore before assembly.
- Using the correct anti-seize lubricant.
- Always replacing sealing washers.
- Carefully observing mounting torque and using the correct/appropriate tool.

If the transducer is water-cooled, switch on the water circulation system before the engine starts and check via sight glasses for the correct water flow.

9.4.4.3 Maintenance and Repair

The maintenance of a pressure transducer generally involves cleaning the outside of the unit (and inside if water cooled) periodically after use, or prior to the start of a measurement program. In particular, attention should be paid during the cleaning process to maintaining high-insulation resistance. This is a critical factor, as it is very important to the reliability and success of the measurement chain as a whole, by reducing drift effects.

Based on the nature of the application, it is likely that the transducer measuring face will become soiled with deposits from the combustion process. This will clearly affect performance of the transducer, as these deposits are poor heat conductors and thus impair the heat flow and reduce the strain caused by it in the diaphragm. The deposits can have a positive or negative effect on the measurement results, depending on the type of pressure transducer used. Soiling of the transducer measuring face can result in an error of up to 10% in the determination of some calculated results.

Therefore some routine cleaning is necessary. This must be carried out with specific care and caution, though; we have already mentioned that transducers are sensitive pieces of equipment. Therefore, use of the correct chemicals and methods is essential. Manufacturer's specific information should always be sought, but here are some general guidelines as to the cleaning of transducers:

- Always ensure that the transducer cable is fully installed and tightened before any cleaning process. This helps to ensure that the electrical connection is sealed and less likely to be contaminated by cleaning fluids (thus reducing insulation resistance).
- Never use any mechanical means of removing deposits from the transducer; abrasives, etc., are to be completely avoided. A typical interim cleaning operation could be wiping the transducer with a soft cloth soaked in an appropriate cleaning solvent (refer to manufacturer instructions for type of solvent to be used).
- A more intensive cleaning could be achieved by soaking the unit in a small quantity of solvent to loosen deposits. Then wipe gently and rinse with clean solvent, and blow dry lightly with clean, compressed air. In addition, cleaning spray recommended by the manufacturer can be used.
- The most extreme, intensive measure is to clean the transducer in an ultrasonic bath. Note that the cable must be installed for this process and that the transducer must not

be completely immersed. Note that a typical cleaning time would be 5 to 10 minutes (no more). Also, check that the power and temperature of the ultrasonic bath, and the cleaning fluid to be used, is appropriate for the transducer. Note that excessive power or cleaning duration will destroy the transducer.

- Attention must also be paid to the cooling water galleries in cooled pressure transducers. If these passages and galleries are clogged with scaling, the heat transfer capability will be reduced, due to restriction of the coolant flow within and around the measuring element. It is possible to clean the transducer with commercially available descalers (amino acid, acetic acid, or formic acid) by circulating the cleaning fluid through the transducer, or leaving it to soak in the fluid to dissolve the scale. An indication of the successful completion of the cleaning process can be gained by monitoring the flow rate through the transducer compared to the characteristic value of a new unit. Once cleaning is complete, rinse with distilled water and dry by blowing out with clean, dry compressed air. Note that this cleaning operation should always be carried out with the unit removed from the measuring position at the engine test bed.

Correct cleaning procedures also form the basis of the scope of transducer repair, what is possible for the user of the transducer. Some small repairs can be carried out to ensure longevity of the transducer: replacing cables, cooling water nipples, and reaming of sealing faces (in the adaptor), for example. In addition, some nonintrusive adaptors (particularly spark plug–type) can be disassembled and have user-replaceable parts—thus, they can be repaired by users. In many cases, though, transducer problems that require repair will need the unit to be sent back to the manufacturer, as they are high-precision components that require micro-precision assembly techniques to disassemble and repair them.

In addition to the transducer itself, cleaning of the cabling and connectors is important to maintain high-insulation resistance. If possible, contamination of cabling should be avoided in the first place; however, it is possible to reinstate high resistance at the cable ($>10^{13}$ ohms) via cleaning of the connectors with appropriate cleaning spray (refer to manufacturer's instructions). Then blow dry with clean, compressed air and then, if necessary, oven warm the cable/connectors to dry them out completely (e.g., 125°C for an hour or so). The insulation resistance can be checked with specific test meters (which are a very worthwhile investment). If the insulation resistance cannot be restored by the above method, the cables and/or connectors must be replaced.

9.4.4.4 Calibration

We have already mentioned that the inputs of the measurement hardware can and generally are calibrated via a traceable reference voltage source, to ensure the accuracy of the data acquisition system and process. Also, all piezoelectric transducers will be fully calibrated by the manufacturer, and often "run in" via operation in a running engine before shipping to the user. In addition, signal conditioning systems will undergo regular calibration checks to ensure their accuracy and repeatability in use. This means that, in theory, connecting all these individually calibrated elements together will form an accurately calibrated measurement chain. In practice, though, this is rarely the case; individual errors of the components can lead to a large summed error. The issues of sensitivity and linearity of the individual components can be difficult to fully account for, and these factors can have

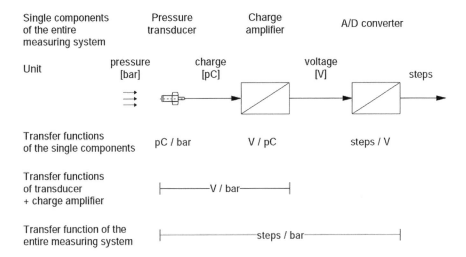

Figure 9.7 Measurement chain calibration.
(Source: AVL.)

cumulative or offsetting effects across the measuring range. For the greatest level of accuracy, it is generally recommended to calibrate the complete measurement chain, as a whole. This is particularly important where a long series of measurements on an important test program will be undertaken. This is a superior approach when compared to standalone calibration of the pressure transducer, followed by inputting its known transfer function into the other components of the measuring system. The latter procedure does not guarantee the same accuracy and certainty as calibrating the entire measuring system.

Many test environments will include calibration equipment, and for the piezoelectric measuring chain, this consists of a dead-weight calibrator. A typical dead-weight calibration system is shown in Figure 9.8. It consists of a hydraulic circuit that applies a known static pressure to the transducer based on a "dead weight" that generates the hydraulic pressure. This pressure is instantaneously released via a valve, which allows the transducer to be subjected to a step decrease in pressure, effectively a square-edge calibration jump which facilitates a 2-point calibration of the transducer, but also the complete measuring chain if connected. The dead-weight calibrator is generally a portable piece of equipment that can be moved and applied to the test system where it is needed *in situ*.

The pressure jump falling edge is normally used for the calibration process, as this is more predictable and repeatable and does not suffer from overshoot; this is shown in Figure 9.9.

Using this method to calibrate the whole measurement chain depends upon this feature being supported in the software or user interface of the measurement device. Most systems do support this method to accurately calibrate the whole system via a single "jump" measurement; this is sufficient, as the linearity deviations of the complete system are generally quite low. However, it is possible to execute a multi-point calibration where highest

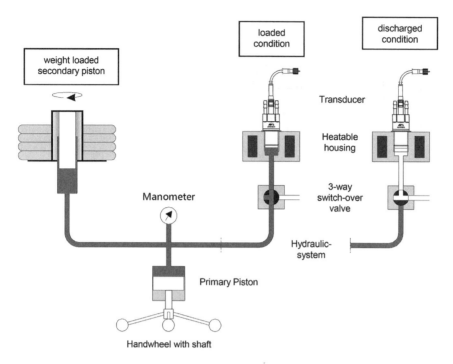

Figure 9.8 Dead-weight tester.
(Source: AVL.)

accuracy is most important. This allows the determination of mean sensitivity and linearity, which than then be compared with the transducer manufacturer's calibration sheet, as this will highlight transducer damage or end of life. The calibration pressure is derived from the following equation:

$$p_K = \frac{F}{A_{SK}} = \frac{g}{A_{SK}} \cdot (M_{SK} + M_Z)$$

F .... plunger force
A_{SK} .... secondary plunger area
M_{SK} .... secondary plunger mass
M_Z .... added mass (weights)
g .... mean gravitational acceleration (9.81 m/s²)

During the measurement of the pressure jump, drift must be minimised by the use of clean, high-quality cable and connectors. This is essential, as any drift compensation circuits in the charge amplifier must be disabled to prevent distortion of the measurement. In addition, optimised use of the analogue-to-digital converter range is important, to ensure that the calibration signal is appropriately digitised with high quality.

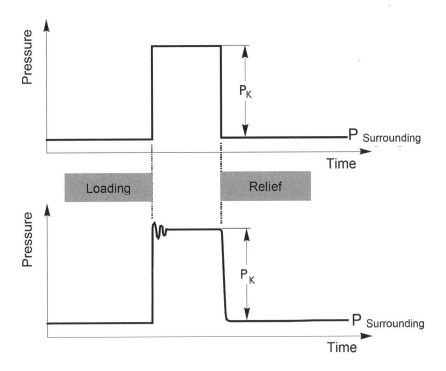

Figure 9.9 Calibration pressure jump.
(Source: AVL.)

The dead-weight tester is simply a flexible method of applying a known pressure to a measurement chain for calibration. The acquisition of the jump, and the interpretation of this into a calibration value for the measured channel, is a function of the measurement system, and therefore, the manufacturer-specific information or documentation should be sourced in order to establish the specific features in the user interface to support calibration. Often, data plausibility functions are included, to ensure that the pressure jump has sufficient quality and dynamics for a high-quality calibration.

In summary, there are some general guidelines to be observed when executing a dead-weight calibration for the complete measurement chain:

- The calibration environment should be as clean as possible. Clean the sensor sealing face before mounting into the dead-weight tester. Check for any leaks in the hydraulic system and replace seals regularly to prevent leakage. Always immediately clean any fluid leakages.
- Use the correct tools for mounting and dismounting sensors. Always apply the correct mounting torques, and handle the transducers with care. And always remember that every time a transducer is mounted and dismounted, the internal stresses in the housing changes, which can affect the overall sensitivity.

- Visually examine the transducer carefully before installation into the dead-weight tester. If possible, also check the insulation resistance.

- Charge amplifier drift compensation circuits must be off. Ensure cleanliness of all cables, connections, and interfaces. Ensure that the analogue input range of the measurement device is fully optimised by appropriate gain setting on the charge amplifier, for maximum conversion quality.

- The calibration pressure is based on the anticipated measurement pressure. If the measurement at the engine is in the lower pressure range, the calibration should also be in this range. In most cases (e.g., IMEP measurement, measurement in high-pressure range), it is appropriate to define the test or calibration pressure at a level of approximately 75% of the expected maximum measurement pressure. For measurement of gas exchange dynamics, 100% of the expected measurement pressure should be chosen. Note, though, that the calibration must always be in the upper half of the measurement pressure range; otherwise the measurement accuracy, particularly in the peak pressure range, will be compromised.

- Ensure that system pressure is fully released from the tester before attempting to dismount the transducer.

Note the following issues that could cause problems when calibrating, specifically related to the dead-weight tester hardware:

- Trapped air in the hydraulic pressure system.
- Nonrotation of the weights, causing error due to "stiction."
- Dead-weight tester must be placed on a level surface.
- The weights must rotate between the upper and lower stop point (i.e., they must be floating).

9.5 Summary

Although the combustion measurement chain is quite specialised hardware—mainly due to the specific requirements of the piezoelectric measurement chain—once it is reasonably well understood, including the specific requirements, it is possible to achieve high-quality measured data with relative ease. A clear understanding of the application is most useful. In addition to the knowledge of the individual parts of the measurement chain and how they work together, as this will facilitate a level of knowledge that can not only support day-to-day operational work. It will also provide a good understanding of errors, why they have occurred, and what can be done about them, either in future measurements or by manipulation of errored data. Of course, normally we don't like to adjust data. Data should be as measured; otherwise it's not representative. However, combustion data is often quite subjective, compared with simulations that are completely subjective. Hence, if the reason for a poor-quality measurement is well understood, then adjustment made in processing can validate the data such that the test bed time is not wasted.

In the past, using combustion measurement systems was the reserve of experienced engineers or scientists, who would often execute the measurement and then process the data. As times and development processes and requirements have changed, the combustion measurement is no longer the preserve of experts; it is a commonly found instrument that is often used on a day-to-day basis. Therefore successfully setting up a system, measuring and storing data, then checking data quality, is no longer confined as a specialist activity that requires years of knowledge and experience. All one really needs is a sound foundation of knowledge and some common sense!

Chapter 10

Specification and Integration into the Test Environment

10.1 Introduction

The combustion measurement system is a complex subsystem that exists as part of the overall test environment, yet plays an essential role in understanding engine behaviour for development and research purposes. This is true because understanding the in-cylinder processes is fundamental to the task of optimising the engine. Many measurement devices exist around the engine test bed, most of them measuring some fundamental input or output from the engine to ensure that it operates within limits, and to understand the engine's consumption and output for the purpose of increasing its efficiency and improving its performance. Most of these measuring systems provide averaged, scalar result values or transient measured curves of fast-sampled points, depending on whether the main focus of the test is steady-state or transient operation of the engine.

The combustion measurement system is, however, quite different: the measurement data are sampled extremely quickly and are related to crankshaft position, producing cycle-based curves of the measured channel against crank position. This information is of interest because most of the engine processes and control subsystems are operating and controlling very dynamic processes, such as initiation of the fuel burn process, which must occur in relation to the correct crank position. These data provide the fundamental understanding of the progress and quality of the combustion process that engine-testing engineers and scientists need.

The real challenge is how to integrate this system effectively into the test environment so that it can be used and operated easily, and so that the system can produce quality data that are accessible and traceable. This can be quite difficult. In the past, the combustion measurement system has been considered an exclusive and expensive tool of a scientist, whereby a typical system would be managed and controlled by an expert who would execute the operation and measurement tasks of that system, then take it away with the data in order to process the data and produce summarised reports. Working in this way, apart from the physical connection to the engine (via sensors and transducers), there is no real integration of the system into the test environment. This scenario is less commonly encountered in modern test facilities. The combustion measurement system is a much more common piece of equipment, often used on a day-to-day basis for measuring and monitoring purposes during various engine test and development procedures. It is therefore much more likely that a test cell or facility engineer will be involved in the task of specifying and purchasing the correct equipment to produce data for an internal or external customer. It may also be the case that the persons buying the equipment may not actually use it, or even understand it.

An additional complication is that the configuration of a combustion measurement system, including all the peripheral equipment, is a complex task—as complicated in some respects as specifying a complete engine test bed. Therefore, some level of knowledge of the equipment and its application is needed in order to correctly specify a system, with appropriate integration into the test environment.

In this chapter, the main features and functions of typical systems are explained. Then typical system configurations are proposed as examples, with suggested applications. In conjunction with the previous chapters, this discussion will assist in developing a sufficient level of basic knowledge of system configuration to be able to specify and propose a complete measuring device for a given environment.

10.2 Technical Considerations—Features and Functions

10.2.1 Introduction

The combustion measurement system consists of a number of subcomponents that form the complete measuring chain. The fundamental pressure measurement from the transducer is at the beginning of the chain, with the data storage and analysis environment at the end of the chain. (Data are the ultimate output from the chain and the system itself.) This model is shown in Figure 10.1 along with the intermediate steps. At each step, some conversion is taking place, either physical or in format.

When a system is being specified or considered for a particular environment, the output required from the chain should be taken into account, and the choice is generally a function of the measurement task or tasks that will be undertaken in the applications. The actual output is data, and the main considerations are where the data will be stored and how they will be accessed and viewed. In addition, how will the data be compared with previous or other measurements? How will the data be synchronised with other data from other systems? These are most important, but often overlooked, considerations.

With respect to overall system components, when specifying, one needs to consider the main functional blocks of the system:

- Transducers. Do these already exist in the test environment? If not, they should be purchased with appropriate cabling and adaptors.
- Encoders. What crank degree resolution is required for the application? Where can the encoder be fitted on the engine? If encoders already exist in the test environment, are they suitable and can they be interfaced?

Figure 10.1 Typical conversion processes for a combustion measurement environment.

- Signal conditioners and cabling. Do they exist already? If so, can they be interfaced to new equipment? How many channels are needed with what type of amplifiers?
- Measurement hardware. Which features and functions are needed? How many channels? How much memory?
- Measurement software and data environment. Which software features are needed? What is included as standard? What is available as an option? How are data handled, measured, and parameterised? Which tools are needed to process and visualise the data?

Of course, this is only a small sample of questions that need to be answered when considering a system configuration for a test environment. A useful approach for the inexperienced is perhaps to look at configurations of some typical systems for common applications, noting the specific features that are of interest in each case.

10.2.2 Typical Systems and Applications

In this subsection, we will consider the attributes of three typical measurement system configurations to fulfill the requirements of commonly found applications. The main differentiating factor with modern combustion measurement systems, with respect to applications, is the channel count. This can be considered as a basic scaling factor for the size and capability of the system because, generally, the more sophisticated the measurement application, the more channels are needed. In addition, as systems increase in channel count, they normally increase in memory and processing power and hence become more powerful and capable systems. For the three examples discussed below, we can categorise them in simple terms as low-end, medium-range, and high-end systems, with typical channels counts and features for the applications suggested.

10.2.2.1 Low-End System

The typical attributes of a low-end system (Figure 10.2) are as follows:

- Four to eight high-speed analogue channels
- Throughput rate of up to 1,000 kHz per input channel
- System memory sufficient to capture up to 1,000 cycles of raw data

Figure 10.2 Typical low-end, compact system packaged with amplifiers.
(Source: AVL.)

- Calculation of standard results from the raw data for display during the measurement
- Small form factor, typically one or two 19-inch rack units
- Lightweight, portable
- DC power supply input possible, sourced from vehicle battery or power supply module
- Combined system and signal conditioning package

Such a system would provide suitable measurement capability for measurement and analysis of the raw data, fulfilling all the requirements of the basic cylinder pressure analysis applications. Generally, the software supplied with all combustion measurement systems will support the specific features needed—for example, correct pegging and top dead centre (TDC) referencing of the pressure trace. In addition, a basic system should be able to calculate the standard results (MEP [mean effective pressure] and heat curve results) for display during the measurement. Typical measurement mode requirements are monitoring mode (i.e., oscilloscope) and single measurement mode; these are initiated by the operator from the software-user interface.

A system of this type is often packaged with the amplifiers to create a portable unit that is shared among a number to test cells—moved to the relevant cell when combustion measurement and data are required (often mounted in a cabinet or trolley). Therefore, the system would most likely be manually operated and the data manually processed after the measurement. Hence, it is often the case that no online automation or data transfer is needed in a system of this type used at an engine test bed.

Often a simple system such as this is used to collect raw data for advanced analysis that will be carried out offline, with the main target being the combustion analysis data. The test bed data is therefore less important, so connection between the two systems is less relevant. A typical application example is a single-cylinder engine test environment, often found where fundamental combustion studies are carried out or in university research facilities. The system does generally not need more than eight channels, and a lower number is possible because in many cases only one cylinder is instrumented (even for multicylinder applications). Typical measured channels are inlet, combustion chamber, and exhaust pressure for a gasoline engine, and combustion pressure, injector, and fuel pressure measurements for a diesel engine. Even if multiple cylinders are instrumented, it is normally just to measure cylinder pressure for each cylinder, giving the four-to-eight-channel requirement.

A system of this type, packaged with amplifiers, will often form the basis of an in-vehicle measurement system, used by engine calibrators for final in-vehicle tuning of an engine calibration. Typically the system is then used with a notebook-type computer for operating the software control interface; this is a compact, self-contained measurement system that can easily be used in mobile measurements applications. The possibility of operation from a low-voltage power source (12-volt DC) then becomes an important feature. In addition, the ability to use an encoder signal from the engine control unit (ECU) electronic pulse wheel becomes an essential feature and capability. For this application, it is common that some software extensions for the combustion measurement system will be needed to allow the system to interface with other onboard data and measurement systems. Two possible scenarios could be encountered: the first is that the combustion measurement system is the master data acquisition system and hence requires interfaces to external devices in order to collect all data centrally (as the master). This often consists of an interface to the engine's electronic control system, where many values are measured or derived anyway for the engine control (and are hence available) and which interfaces with data-acquisition modules for measuring additional channels (temperatures and pressures from nonproduction sensors). Centrally collecting these data in the combustion measurement system allows appropriate correlation between the crank angle data, cyclic results, and additional, time-based measured values. The second scenario is that the combustion measurement system calculates results and outputs them continuously over an interface where they can be acquired by another measurement system. With this method, though, all the acquired data are generally aligned as time-based data. Interface requirements for the former could be ASAP-3 for ECU data (or other standardised interface for application systems). For the latter scenario, commonly used interface technology is controller area networking (CAN), USB, or TCP/IP. Additional measurement functions needed in the control software of the combustion measurement system likely for in-vehicle work could be cold-start measurement, knock, or combustion noise. These can also be purchased at a later date if needed. In summary, a low-end system is a basic combustion analyser in a small, flexible package that is used to collect data at the test bed, or in the vehicle, mostly for post-measurement analysis. The channel count and processing performance are not high requirements; low cost and flexibility are more important.

10.2.2.2 Mid-Range System

The typical attributes of a mid-range system (Figure 10.3) are the following:

- 8–16 high-speed analogue channels
- Throughput rate of up to 1,000 kHz per input channel

Figure 10.3 Mid range system mounted permanently at a test bed system
(Source: AVL.)

- System memory sufficient to capture up to 1,000–3,000 cycles of raw data
- Calculation of standard results from the raw data for display during the measurement; also, the ability to adjust standard algorithms and implement self-developed calculations
- Compact form factor, typically two or four 19-inch rack units
- Besides standard combustion measurement functions, additional options according to the application, such as knock measurement, cold start, simultaneous time based acquisition
- Remote control interface to test bed automation system/host

A system of this type may be used as a standard day-to-day measurement system for combustion data, supporting research and development activities such as engine calibration. Typically, 8 to 16 channels are needed to support multicylinder engines, with multiple channels per cylinder required. Throughput rate and memory, in addition to processing performance, should scale up with the channel count. This is often the case anyway, as most systems are modular with acquisition cards that support a certain number of channels; each card has its own processor and memory, so increasing the number of channels increases the required card count and overall system performance.

A system of this form is often permanently installed at the test bed as part of the environment, but may also be installed in a trolley arrangement as a portable system to be moved around the test facility. When mounted in a trolley, appropriate interface panels for easy connections should be considered to reduced rigging and setup time when the system is relocated.

A remote control and data transfer interface is likely to be required for the mid-range system. Remote control of the measurements during the automatic test run is a feature that will improve test bed runtime efficiency. In addition, the ability to transfer data to the test bed data model facilitates a more productive data environment for access and post-processing. For certain activities, a reliable connection between the combustion measurement and test bed automation system is an essential prerequisite—for example, engine calibration environments where calculated results for knock and burn rates are essential for the task—in addition to being needed for engine protection.

Additional software options may be needed according to the required activities at the test environment; these could include the functions for measuring knock, noise, cold-start, or torsional vibration. In addition, the ability to execute self-developed combustion calculations and algorithms may be a needed feature.

Most mid-range systems append the analogue input channels with digital inputs and outputs, and these may be useful for simple, trigger interfaces for alarm or warning purposes. A possible example is the triggering of a digital output change of state if maximum cylinder pressure exceeds a certain level (to provide engine protection).

In summary, mid range systems are used for day-to-day measurement tasks, mostly at the engine test bed. They are capable of standard measurement and result calculation, and they have the capability to provide an interface between themselves and other systems for data transfer and control, which is often required for many current measurement applications. The system hardware is generally capable of more advanced measurement and analysis tasks, although this may necessitate the purchase of additional, nonstandard options from the supplier.

10.2.2.3 High-End System

The typical attributes of a high-end combustion measurement system (Figure 10.4) are as follows:

- 16–32 high-speed analogue channels, possibly possible by scaling up a modular system
- Throughput rate of up to 1,000 kHz per input channel

Figure 10.4 A high-end system, trolley mounted
(Source: AVL.)

- System memory sufficient to capture up to many thousands of cycles of raw data
- Calculation of standard results from the raw data for display during the measurement, and transfer through external interfaces to other measurement systems
- The ability to adjust standard algorithms and implement self-developed calculations by scripting interfaces for calculations and measurement control
- Typically three or four 19-inch rack units, with multiple units cascadeable to increase overall channel count easily
- Besides standard combustion measurement functions, additional options included for advanced measurement and analysis: knock measurement, cold start, combustion noise, advanced filtering, torsion and vibration analysis, and simultaneous time-based acquisition
- Remote control interface to test bed automation system/host; fast data transfer outputs for external logging or results, or for use in control loops

A system of this type can be employed for the most demanding measurement applications, where high-speed logging and processing performance in real time are the most essential attributes required of the system. It can also be employed for applications in which a high channel count is needed for simultaneous crank angle and time-based high-speed logging. Real-time performance is essential, and a common requirement in high-performance applications is the ability to transfer results to external logging or measurement systems in real time. This transfer can be accomplished through a hybrid interface (i.e., a combination of digital and analogue outputs) or through a high-speed digital interface such as CAN. Whichever interface is used, the performance of the system in transferring data values is essential. There are two main areas of system performance: (1) the internal processing speed for calculating the results in real time, and (2) the interface performance in making those results available to other systems. The high-end system should provide the necessary performance in both these areas. Channel count of a standard system is 16 to 32 channels, but this should be extendable via additional cards (as in the mid-range system). In addition, though, high-end systems are often capable of extending channels with additional data acquisition units (i.e., the racks in which the cards are housed) that can be connected, thus facilitating a system that can be cascaded up to nearly 100 high-speed channels if required. In addition, high-end systems generally accommodate simultaneous high-speed acquisition of time-based data; the combustion measurement system thus becomes a complete high-speed logging environment that allows time- and angle-based data to be acquired, visualised, and stored in real time. Often, digital timers and counters are also supported in high-end systems that allow additional advanced measurement features such as instantaneous speed calculation from rotary encoders, so that torsional and rotation analysis can be performed, along with pulse width and modulation analysis.

Typical applications for high-end systems are development environments for high-performance engines (e.g., motor sport and racing engines) because these environments generally need fast processing and output of results. Other examples include advanced development environments where multicylinder engines, with advanced combustion and control systems, are being developed or calibrated. High-end systems should support the future requirements of alternative in-cylinder-based measurements (e.g., combustion visualisation) in addition to standard pressure-based analysis. These environments require a high-performance measurement system with uncompromised data throughput and the ability to calculate results quickly and store large data files.

In summary, a high-end system is often needed where large channels counts or high system performance/throughput is needed. The system should have all the performance of the mid-range system, with greater and extendable channel count. It should include flexible, high-performance options for transferring data and all the additional features and functions for advanced measurements within the standard equipment supply (which are only options on the mid-range system).

10.3 Interfaces to Additional Equipment

10.3.1 Introduction

Interfaces are an important consideration for the combustion measurement system. As has been discussed throughout this book, it is often necessary to transfer data or implement control functions between the combustion measurement system and external measurement or control systems. In addition, it is often the case that the various interconnecting parts of the measurement system may not be from a single source, and this then necessitates some thought and pre-planning of how components will connect physically—but, more importantly, will they work properly when connected?

Interfacing is one of the most crucial factors to consider when purchasing, installing, or integrating a combustion measurement chain, yet it is often implemented as an afterthought and only studied when the system does not work properly. This subsection will highlight the main points to consider when specifying equipment for the measurement chain, and when considering integration of the combustion measurement system into the test environment.

10.3.2 Interfaces—Measurement Chain Components

The combustion measurement system consists of a number of key components that have been discussed in this book. It is crucial that they all work in a harmonised way, but they are often provided by different suppliers. Many manufacturers of equipment specialise only in certain parts of the system as their key area of competence. Hence, in a working environment the key components are unlikely to be from the same supplier. Also consider that, even if they are from the same supplier, they may be different generations of equipment, so compatibility is not guaranteed! The main interface points between the components are discussed below, with emphasis on the areas of concern that should be considered when purchasing or integrating equipment.

10.3.2.1 Angle Encoder to Measurement System

This is one of the more difficult interfaces to implement properly: many different types of generic encoder on the market may be encountered, and there are several manufacturers of equipment specific to the application. Therefore, a mismatch between encoder signal and measurement system may often occur. Generally speaking, a combustion measurement system is operated via two signals: a once-per-revolution trigger, and a number of crank degree marks per revolution. Almost exclusively, the signals are square wave pulses. Interference on the signal transmission lines can cause false triggers, and excessive inductance or capacitance on the cabling can cause poor-quality signals with undefined signal edges that cannot be interpreted by the measurement system. In general terms, one should ensure that the specifications of the encoder input requirement are known and fulfilled. Differential signal transmission can often be used to reduce noise, but the encoder and the measurement system should have matching interfaces to support this. If possible, choose an encoder that is ground isolated so that any ground potential difference does not circulate through the encoder cabling. Ensure that encoder hardware setting is correct (pull-up resistors, internal pulse multipliers) so the correct number of active edges is provided to the measurement system. Ensure that the correct edge is applied for the triggering and that, if necessary, the trigger and crank degree mark active edges are appropriately misaligned so they can be interpreted and discriminated correctly. Ensure that voltage-level outputs are correctly applied (TTL, CMOS, or other).

10.3.2.2 Transducer to Charge Amplifier

The main issue with this interface is to check that the amplifier and transducer are harmonised with respect to grounding: the system technology should support immunity to ground loops by isolating parts of the measurement chain from each other. This isolation is generally accomplished with differential inputs or outputs (or both) at the charge amplifier, or with ground-isolated sensors—but not both. This aspect should therefore be checked, or one or the other approach should be adopted as the standard approach within a test environment. Applying this convention will help to prevent a mismatch that will result in an incomplete circuit, which will stop correct operation.

Another aspect to consider at this interface is the measurement cabling. The correct type of high-impedance cabling should always be used, one that is resistant to triboelectric effects. In addition, the cable should be laid carefully, off the ground to prevent contamination from dirt and grease. The cable must be fixed to prevent movement from vibration transmitted by a running engine.

10.3.2.3 Charge Amplifier to Measurement System

Most systems interface through an analogue voltage applied directly to the measurement system's high-speed channel input with analogue cabling (typically BNC connectors). These cables should be routed carefully and appropriately. The foremost consideration is to ensure that the charge amplifier settings are set to maximise the dynamic input range of the measurement device. Often the output from the amplifier and the input to the measuring device are ground-isolated differential inputs to reduce noise and grounding issues; however, to prevent the possibility of further ground and noise issues, the measurement device and the amplifier rack must both be connected to the main instrument ground via substantial earth-bonding cables, because this will help to prevent interference due to ground loop issues. (See Figure 10.5.)

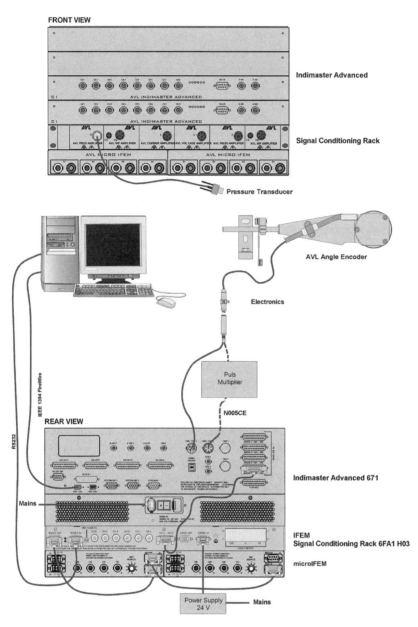

Figure 10.5 Typical system overview showing interfaces
(Source: AVL.)

10.3.3 Interfaces—Data Transfer and Control

As mentioned previously in this book, the combustion measurement system generally has interfaces to external devices or systems for control and data transfer. These are often required for the overall development task, to control the combustion measurements and to save or transfer data to the test bed controller or host system. There are a number of possibilities for this integration, two of which are discussed here.

10.3.3.1 Digital Control and Data Transfer Interface

This type of interface is frequently used and required. In fact, many manufacturers of combustion measurement systems also have a product portfolio that includes control systems for test automation. It is a logical extension for a manufacturer to offer the possibility of connecting the devices. When this task is executed via an intelligent bidirectional digital interface, the combustion measurement system becomes a "slave" measurement device. Its operation, including measurement, loading parameters, storing raw data, and transferring statistics or result values from the measurement, then becomes a function of the test bed control interface. This setup is preferable because the operator of the test bed can then control the combustion measurement system in a way similar to other measurement devices, because measurements can be initiated during an automatic test run (automatically!), and because the result data from the combustion measurement device are integrated with the test bed data model, also being viewed on the same screen.

An important point to note with this type of interface is that generally the bandwidth for result data transfer is not particularly high: the overheads of the protocols that include bidirectional data result in a transfer rate of actual data that is quite low—as low 1 Hz depending upon the protocol being used. Therefore, unless the data transfer requirement is low during online operation, the main function of this interface should be considered as remote control of the combustion measurement system. Typically, the most appropriate application for data transfer on this interface is to transfer statistics of result values that have been measured at a steady state log point, because continuous transfer of all results is generally too slow. The technology is improved with the use of standard computer network technology (TCP/IP) between the automation and combustion measurement system, and this improves performance by an order of magnitude (up to 10 Hz), still too slow for many applications. There are other alternatives (discussed below) that can provide higher data throughput.

An important point to note with these interfaces is that there is no standard definition; each manufacturer has its own protocol running on the interface (normally RS232 or TCP/IP). Many systems offer flexible possibilities so that users can write their own interface as the command set is supplied by the manufacturer. A word of warning, though: when systems from different manufacturers are connected, it is often the case that some features or commands in the interface will not be implemented. In addition, any slight error in the program command syntax will cause the interface to fail. Hence, when these interfaces are connected, they are very rarely "plug-and-play"; it is quite rare that the connections work at the first attempt, and time should always be allowed, on first connection, for bug fixing. Make sure that development resources are also available at this time.

10.3.3.2 Analogue and Digital Hybrid Interface

This type of interface does not allow any intelligent control of the measurement device: it simply allows the possibility to send data values, in real time, from the combustion measurement system to an external measurement device. A common feature is the digital-to-analogue converter, which is an extra internal component in the measurement hardware that facilitates the continuous output of result values as analogue voltages. The combustion measurement system allows scaling of the voltage output such that it is in the correct range relative to the result (e.g., 0–10 bar IMEP = 0–10 volts). This output can then be logged on an external device or measurement system. The advantage is that the analogue output is fast reacting. In fact, in most cases it is real-time output, so critical values used in control or optimisation loops can be logged and displayed deterministically at the test bed. The disadvantage of this approach is that summed errors when scaling inputs and outputs will reduce overall accuracy of the measured values when transferred to the test bed or host. In addition, there is normally a limit on the number of outputs (often 16), which restricts the number of results transferable.

The digital interface to the combustion measurement system allows input/output of digital signals that can be used to trigger measurements or other simple tasks (e.g., saving data). In addition, alarm signals can set up as output based on measured combustion result values that exceed predefined limits. This allows a limited amount of automation and communication that is simple to set up.

However, the disadvantage for the analogue and digital interface is the additional physical wiring and cabling effort needed to integrate the signals between systems. This effort, often quite time-consuming, should not be underestimated.

An interesting recent development for transferring results quickly to other systems is the adoption of the use of CAN (controller area network) technology for this purpose. CAN is a standardised interface in the automotive industry that is almost universally adopted in vehicle wiring systems for data exchange between control units. The main features of CAN are its high reliability and safety with respect to data transfer of critical values. It is a robust and proven interface with high levels of redundancy and fault tolerance, and it is therefore ideal for the task. In fact, CAN is adopted in several other areas of measurement technologies outside the automotive industry (medical and process industries at the time of writing). The CAN protocol allows 64 bits of data for transfer of digital or analogue values. In addition, messages can be prioritised via a message identifier, because the CAN network is a multi-master system that can support many nodes on a single bus. Each node has the same access to the bus, and collisions are handled by the priority definition of the message.

The combustion measurement system may include the required CAN hardware to write the result values into individual messages that are then transmitted on the bus. Often a separate CAN network between the combustion measurement and test bed automation system provides the best performance and throughput, with a capability to transfer data at a maximum of 1 MBit, which is the fastest speed that a CAN bus currently allows (at the time of writing). The advantage of CAN—apart from that fact that is standard, well documented, and understood in the industry—is that one CAN network (i.e., one cable) is capable of transferring hundreds of results at very high speed. This therefore allows fast, pseudo-real-time data transfer of results with a very simple physical cabling requirement between systems. Also, digitally transferred data will not suffer the scaling inaccuracy of analogue inputs and outputs.

Note that at the time of writing, though, CAN interfaces for combustion measurement systems are used only for transferring data; no control through this interface has yet been proposed by any manufacturer.

Bibliography

AVL List GmBH. *Engine Indicating User Handbook.* Graz: AVL List GmBH, 2002.

AVL List GmBH. *Engine Indicating with Piezoelectric Transducers.* Graz: AVL List GmBH, January 1996.

AVL List GmBH. *Flame Measurement Techniques for Engine Development Engineers.* Graz: AVL List GmBH, 2007.

Basshuysen, R. and F. Schaefer. *Internal Combustion Engine Handbook.* Warrendale, PA: SAE, 2004.

Eastop, T. D. and A. McConkey. *Applied Thermodynamics for Engineering Technologists,* 5th ed. Essex, UK: Longman Group, 1995.

Gossweiler, C. R., W. Sailer, and C. Cater. *Sensors and Amplifiers for Combustion Analysis.* Winterthur: Kistler Instruments AG, 2005.

Greene, A. B. and G. G. Lucas. *The Testing of Internal Combustion Engines.* Bucks, UK: English University Press, 1969.

Heisler, H. *Advanced Engine Technology.* Warrendale, PA: SAE, 1995.

Heywood, J. B. *Internal Combustion Engine Fundamentals.* Columbus: McGraw-Hill, 1988.

Martyr, A. J. and M. A. Plint. *Engine Testing,* 3rd ed. Oxford: Elsevier, 2007.

Robert Bosch GmBH. *Automotive Handbook,* 6th ed. Westminster, UK: Copublished Professional Engineering Publications, 2004.

Stone, R. *Introduction to Internal Combustion Engine Fundamentals,* 3rd ed. Hampshire, UK: Palgrave, 1999.

Stone, R. and J. K. Ball. *Automotive Engineering Fundamentals.* Warrendale, PA: SAE, 2004.

Zhao, H. and N. Ladommatos. *Engine Combustion Instrumentation and Diagnostics.* Warrendale, PA: SAE, 2001.

References

1. Dobbie McInnes Ltd and Dobbie McInnes & Clyde Ltd, not dated, various editions. *The Engine Indicator: Diesel, Steam.* (With Specimen Diagrams Showing Engine Faults). Glasgow: Dobbie McInnes Ltd and Dobbie McInnes & Clyde Ltd.

2. Trill Indicator Company, 1916. *The Indicator Book* (A Description of the Latest Types of Engine Indicators and Accessories, with Information on Taking and Reading of Indicator Cards). Corry, Pennsylvania: Trill Indicator Company.

3. The American Society of Mechanical Engineers, 1970 (reaffirmed, 1985). *Measurement of Indicated Power* (Part 8, "Instruments and Apparatus," Supplement to ASME Performance Test Codes). New York: The American Society of Mechanical Engineers.

4. Charles Day. Wh.Sc., M.I.Mech.E, 1898. *Indicator Diagrams and Engine and Boiler Testing*, third edition. Manchester: The Technical Publishing Co. Ltd.

5. J. Okill, 1938. *Autographic Indicators for Internal Combustion Engines.* London: Edward Arnold & Company.

6. Thomas J. Main and Thomas Brown, second edition, 1850; fifth edition, 1864. *The Indicator and Dynamometer* (With Their Practical Applications to the Steam Engine). Portsea, Hampshire: Herbert, London, and Woodward.

7. M. F. Russell, 1982. *Diesel Engine Noise: Control at Source* (SAE paper No. 820238). Warrendale, PA: Society of Automotive Engineers.

8. M. F. Russell and R. Haworth, 1985. *Combustion Noise from High Speed Direction Injection Diesel Engines* (SAE Paper No. 850973). Warrendale, PA: Society of Automotive Engineers.

9. M. F. Russell and C.D. Young, 1985. *Measurement of Diesel Combustion Noise* (Proceedings of IMechE Conference). Bloomfield Hills, MI: Autotech.

10. A. E. W. Austen and T. Priede, 1958. *Origins of Diesel Engine Noise* (Symposium on Noise and Its Suppression). London: Institution of Mechanical Engineers.

11. A. J. Herbert and M. F. Russell, 2009. "Measurement of Combustion Noise in Diesel Engines," in *Drivven Combustion Analysis Toolkit (DCAT): User's Manual.* San Antonio: Mechanical Engineering Technology.

12. T. E. Reinhart, 1987. *An Evaluation of the Lucas Combustion Noise Meter on Cummins B Series Engine* (SAE Paper No. 870952). Warrendale, PA: Society of Automotive Engineers.

13. M. D. Checkel and J. D. Dale, 1986. *Computerised Knock Detection from Engine Pressure Records* (SAE Paper No. 860028). Warrendale, PA: Society of Automotive Engineers.

14. A. L. Randolph, 1990. *Methods of Processing Cylinder Pressure Transducer Signals to*

Maximize Accuracy (SAE Paper No. 900170). Warrendale, PA: Society of Automotive Engineers.

15. M. F. Brunt, C. R. Pond, and J. Biundo, 1998. *Gasoline Engine Knock Analysis Using Cylinder Pressure Data* (SAE Paper No. 980896). Warrendale, PA: Society of Automotive Engineers.

16. M. F. Brunt and C. R. Pond, 1997. *Evaluation Techniques for Absolute Cylinder Pressure Correction* (SAE Paper No. 970036). Warrendale, PA: Society of Automotive Engineers.

17. M. F. Brunt and A. L. Emtage, 1997. *Evaluation of Burn Rate Routines and Analysis Errors* (SAE Paper No. 970037). Warrendale, PA: Society of Automotive Engineers.

18. M. F. Brunt and A. L. Emtage, 1998. *Calculation of Heat Release Energy from Engine Cylinder Pressure Data* (SAE Paper No. 981052) Warrendale, PA: Society of Automotive Engineers.

19. M. D. Checkel and J. D. Dale, 1986. *Testing a Third Derivative Knock Indicator on a Production Engine* (SAE Paper No. 861216). Warrendale, PA: Society of Automotive Engineers.

20. D. R. Lancaster, R. B. Krieger, and J. H. Lienesch, 1976. *Measurement and Analysis of Engine Pressure Data* (SAE Paper No. 750026). Warrendale, PA: Society of Automotive Engineers.

21. M. D. Checkel and J. D. Dale, 1989. *Pressure Trace Knock Measurements in Current SI Production Engines* (SAE Paper No. 890243). Warrendale, PA: Society of Automotive Engineers.

22. M. F. Brunt and A. L. Emtage, 1996. *Evaluation of IMEP Routines and Analysis Errors* (SAE Paper No. 960609). Warrendale, PA: Society of Automotive Engineers.

23. G. M. Rassweiler and L. Withrow, 1938. *Motion Pictures of Engine Flames Correlated with Pressure Cards* (SAE Transactions, Vol 42). Warrendale, PA: Society of Automotive Engineers.

24. W. L. Brown, 1967. *Methods for Evaluating Requirements and Errors in Cylinder Pressure Measurements* (SAE Paper No. 670008). Warrendale, PA: Society of Automotive Engineers.

25. G. A. Karim and M. O. Khan, 1971. *An Examination of Some of the Errors Normally Associated with the Calculation of Apparent Rates of Combustion Heat Release in Engines* (SAE Paper No. 710135). Warrendale, PA: Society of Automotive Engineers.

26. R. S. Davis and G. J. Patterson, 2006. *Cylinder Pressure Data Quality Checks and Procedures to Maximize Data Accuracy* (SAE Paper No. 2006-01-1346). Warrendale, PA: Society of Automotive Engineers.

27. M. J. Stas, 1996. *Thermodynamic Determination of TDC in Piston Combustion Engines* (SAE Paper No. 960610). Warrendale, PA: Society of Automotive Engineers.

28. M. Morishita and T. Kushiyama, 1997. *An Improved Method of Determining the TDC Position in a PV-Diagram* (SAE Paper No. 970062). Warrendale, PA: Society of Automotive Engineers.

29. K. S. Kim and S. S. Kim, 1989. *Measurement of Dynamic TDC in SI Engines Using Microwave Sensor, Proximity Probe, and Pressure Transducers* (SAE Paper No. 891823). Warrendale, PA: Society of Automotive Engineers.

30. M. F. Brunt and G. G. Lucas, 1991. *The Effect of Crank Angle Resolution on Cylinder Pressure Analysis* (SAE Paper No. 910041). Warrendale, PA: Society of Automotive Engineers.

31. C. Burkhardt, M. Gnielka, C. Gossweiler, D. Karst, M. Schnepf, J. von Berg, and P. Wolfer, 2003. *Optimization of Gas Exchange by a Suitable Combination of Pressure Indicating, Analysis and Simulation.* (9th Symposium on the Working Process of the Internal Combustion Engine). Graz: Institute for Internal Combustion Engines and Thermodynamics, Graz University of Technology.

32. R. H. Kuratle, 1995. *The State of the Art Combustion Engine Pressure Instrumentation.* Winterthur: Kistler Instruments A.G.

33. R. H. Kuratle and M. Balz, 1992. *Influencing Parameters and Error Sources during Indication on Internal Combustion Engines* (SAE Paper No. 920233). Warrendale, PA: Society of Automotive Engineers.

34. A. L. Randolph, 1990. *Cylinder Pressure Transducer Mounting Techniques to Maximize Data Accuracy* (SAE Paper No. 900171). Warrendale, PA: Society of Automotive Engineers.

33. R. H. Kuratle and A. Signer, 1999. *The Basics of Piezoelectric Measurement Technology.* Winterthur: Kistler Instruments A.G.

34. B. Mandel, P. Wolfer, S. Brechbuhl, and M. Stockli, 1996. *New Indicating Technology on Modern Diesel Engines with Direct Injection Systems and Significantly Higher Injection Pressures* (Common Rail). Winterthur: Kistler Instruments A.G.

35. M. Glavmo, P. Spadafora, R. Bosch, 1999. *Closed Loop Start of Combustion Control Utilizing Ionization Sensing in a Diesel Engine* (SAE Paper No. 1999-01-0549). Warrendale, PA: Society of Automotive Engineers.

36. E. N. Balles, E. A. Van Dyne, A. M. Wahl, K. Ratton, and Lai Ming-Chia, 1998. *In-Cylinder Air/Fuel Ratio Approximation Using Spark Gap Ionization Sensing* (SAE Paper No. 980166). Warrendale, PA: Society of Automotive Engineers.

37. R. Reinmann, A. Saitzkoff, B. Lassesson, and P. Strandh, 1998. *Fuel and Additive Influence on the Ion Current* (SAE Paper No. 980161). Warrendale, PA: Society of Automotive Engineers.

38. M. Hellring and U. Holmberg, 1998. *An Ion Current Based Peak-Finding Algorithm for Pressure Peak Position Estimation* (SAE Paper No. 00FL-587). Warrendale, PA: Society of

Automotive Engineers.

39. M. Asano, T. Kuma, K. Kajitani, and M. Takeuchi, 1998. *Development of a New Ion Current Combustion Control System* (SAE Paper No. 980162). Warrendale, PA: Society of Automotive Engineers.

40. C. F. Daniels, G. G. Zhu, and J. Winkelman, 2003. *Inaudible Knock and Partial-Burn Detection Using In-Cylinder Ionization Signal* (SAE Paper No. 2003-01-3149). Warrendale, PA: Society of Automotive Engineers.

41. E. Winklhofer, C. Beidl, H. Philipp, and W. F. Piock, 2001. *Micro-Optic Sensor Techniques for Flame Diagnostics* (Paper No. 20015301, J., SAE Spring Convention). Warrendale, PA: Society of Automotive Engineers.

42. E. Winklhofer, H. Nohira, C. Beidl, A. Hirsch, and W. F. Piock, 2004. *Combustion Quality Assessment for New Generation Gasoline Engines* (Paper No. 20045451, J). Warrendale, PA: Society of Automotive Engineers.

43. M. T. Wlodarczyk, 2005. *Miniature Fiber Optics-Based Cylinder Pressure Sensors for Advanced Engines* (Paper No. PTNSS P05-C146, PTNSS Congress). Krackow, Poland: PTNSS.

44. A. D. Kurtz, A. Kane, S. Goodman, W. Landmann, L. Geras, and A. A. Ned, 2004. *High Accuracy Piezoresistive Internal Combustion Engine Transducers* (Presented at the Automotive Testing Expo). Leonia, NJ: Kulite Semiconductor Products, Inc.

45. H. Philipp, A. Hirsch, M. Baumgartner, G. Fernitz, C. Beidl, W. F. Piock, and E. Winklhofer, 2001. *Localization of Knock Events in Direct Injection Gasoline Engines* (SAE Paper No. 2001-01-1199). Warrendale, PA: Society of Automotive Engineers.

Index

Note: page numbers followed by an f indicate a figure; page numbers followed by a t indicate a table.

Abnormal combustion. *See* Cold start measurement; Combustion knock; Combustion noise
Amplifiers, 112–113, 113f
   carrier-frequency amplifier, 122–124, 123f, 124f
   charge amplifiers
      basic function and operation, 113
      cabling and interfaces to, 117
      design, 113
      drift and drift compensation, 116–117, 116f, 117f
      measurement procedure, 113–114
      piezoelectric sensor, 113
      time constant property, 115–116
   intelligent amplifiers
      advantages, 125–126
      analogue, 124
      for combustion noise, 133–134
      digital, 125, 125f
      IMEP (indicated mean effective pressure) meter, 135–136
      knock meter, 134–135
      PMax monitoring, 131–133, 133f, 134f, 135f
      properties, 124
      SID (sensor identification), 130, 130f
      Transducer Electronic Data Sheet (TEDS), 126–129, 126f, 127f, 128f, 129f
Analogue signals
   inputs and outputs, combustion measurement system hardware, 139–142, 140f
   mechanical operation, 118–119
   types of signals, 119, 120f, 121
Analysis of diagrams
   measurement process, 12–13
   mechanical operation, 13
   planimeter, 12f, 13
Angle encoder
   basic function, 22
   and combustion pressure curve
      definition, 23
      influence of crank angle resolution on, 24f, 24t
      measurement process, 23–24
      sampling frequency, 25
   inductive system, 22–23
   mounted on the front pulley, 22f
   optical system, 22
   output signal, 25–26
      measurement procedure, 25–26
      mechanical procedure, 25
      profile, typical, 26f
   principles, optical and inductive, 23f
   processing, 30–33
      accuracy of, 33
      and crank position sensor, 32f
      and crankshaft position information, 30
      limitations, 32
      measurement process, 32
      stages of testing, 31
      and toothed wheel, 30–31
      waveform display, 31f
   resolution required, 23–25
   types, 27–29
ASAM. *See* Association for Standardisation of Automation and Measuring Systems (ASAM)
Association for Standardisation of Automation and Measuring Systems (ASAM), 149
AVL Company, 130

Bedel, Charles, 6
Bosch Company, 101

Calculations and results, measurement system software
   derived results, importance of, 163
   future developments, 169–170
   real-time results, 164–165
   user-defined results, 165–169, 167f, 168f
Calibration, transducers, 280–284, 281f, 282f, 283f
Capacitive probe
   advantage of, 191–192
   installation, 195–196
   measurement procedure, 193–194
   mechanical operation, 191
   sensor, 192f, 193f, 194f, 195f, 196f
Carrier-frequency amplifier, 122–124, 123f, 124f
Cetane Index, 101
Charge amplifier
   basic function and operation, 113
   cabling and interfaces to, 117
   design, 113
   drift and drift compensation, 116–117, 116f, 117f
   measurement procedure, 113–114
   piezoelectric sensor, 113
   time constant property, 115–116
Cold start measurement
   definition, 249–250
   significance of, 249
   system configuration and results
      data visualization and processing, 253, 254f
      measurement procedure, 252–253
      preconditions, 250–251
      and quality considerations, 253–254
Combustion knock
   and algorithm adaptations, 241
   causes, 219–220
   conditions producing, 221
   confusion about, 218
   cylinder pressure measurements
      acquisition frequency, 225–226
      acquisition window, 226

      AVL histogram, 234–236, 235f
      AVL KI (knock index), 233, 233f
      AVL real time, 228–231, 230f
      AVL transient, 234
      calculation of knock overpressure, 226, 227f, 228
      FEV CAS, 231–232, 233f
      filtering, 224–225
      frequency, 224
      oscillation, 223f
      rectification of peak curve, 228, 229f
      signal processing, 224
      third derivative, 236–237
      weighting table, 232f
   definition, 218–219, 219f
   measurement techniques
      instrumented spark plug, 222–223
      ionization current, 222
      optical access, 222
      structure-bourne knock sensors, 221–222
   quality considerations
      measurement range and resolution, 239
      measurement system setup, 239–240
      transducer position, type, and properties, 237–238, 238f, 239f
   significance of, 220–221
Combustion measurement system, applications
   chain properties, introduction to, 176–177
   instrumentation
      angle domain, 177
      angle encoders, 177
      external interfaces, 180
      hardware, 178–179
      pressure, 177–178
      signal conditioning, 178
   low pressure–gas exchange process
      and low pressure transducers, 214–216, 215f
      measurement, analysis, 213–214, 214f
      measurement, setup, 212–213, 213f
      measurement, task and goal, 212–212
      mechanical operation, 209–210

TDC (top dead centre)
  capacitive probe, 191–196. *See also* Capacitive probe
  determination methods, 197–200, 199f
  interval method, 190f
  introduction, 185–186, 186f
  microwave, 197
  motored curve method, 191
  pressure curve determination, 188–191
  static determination, 186–187, 187f
  thermodynamic loss angle, 187f, 188f
  thermodynamic analysis
    calculations, offline, 206–207, 207f
    definition, 200–201
    error-producing effects, 208–209
    heat release algorithm, 203–206, 204f
    measurement procedure, 203–206
    mechanical operation, 202–203
  zero-level correction
    detection methods, 184f
    fixed point and measured value, 180–181, 181f
    fixed point and reference value, 179–180, 180f
    introduction, 179
    post-processing, 183–184
    thermodynamic, 181–183, 182f
Combustion measurement system, errors
  quality considerations, 256
  sources of
    amplifier, 259–260
    cabling, 257–258
    encoder, 258–259
    and measurement chain, 260–261, 261t
    measurement device, 260
    transducer, 257
Combustion measurement system, hardware
  measurement data-flow sequence, 137–139
  operating requirements, 137
  system interfaces
    analogue inputs and outputs, 139–142
    angle encoder, 139
    digital inputs and outputs, 142
    operator interface, 139
    typical system, 142–144, 143f
Combustion measurement system, maintenance
  encoder, 276–278, 277t
  hardware and system, 275–276
  transducers
    calibration, 280–284, 281f, 282f, 283f
    installation and handling, 278–279
    maintenance and repair, 279–280
    quality considerations, 278
Combustion measurement system, modern
  angle encoder, 20
  data acquisition system, 20
  dead weight tester, 20
  measurement procedure, 19
  signal conditioning amplifier, 19–20
  TDC (engine tope dead centre) sensor equipment, 20
  transducer, 19
  transducer conditioning system, 20
  workflow process, 19f
Combustion measurement system, setup
  basic, 261–262
  diagnostic and reference measurements, 265–268, 265f, 266f, 267f
  prechecks
    amplifiers, 263
    encoder, 263–264
    measurement procedure, 264
    measurement setup, 262
    transducer and cabling, 262–263
  software and data handling
    best practices, 274–275
    measured data requirements, 269–272
    plausibility checks and data validation, 272–273, 273f, 274f
    reliability, 269
Combustion measurement system, signal conditioning
  and amplifier systems, 112–113, 113f
  analogue signals
    mechanical operation, 118–119
    types of signals, 119, 120f, 121

Combustion measurement system, signal conditioning *(continued)*
  carrier-frequency amplifier, 122–124, 123f, 124f
  charge amplifier
    basic function and operation, 113
    cabling and interfaces to, 117
    design, 113, 115f
    drift and drift compensation, 116–117, 116f, 117f
    measurement procedure, 113–114
    piezoelectric sensor, 113
    and time constant property, 115
  design, 112, 113f, 114f
  harmonisation of elements, 131
  ignition timing amplifier, 121–122, 122f
  information provided, 130
  intelligent amplifiers
    advantages, 125–126
    analogue, 124
    for combustion noise, 133–134
    digital, 125, 125f
    IMEP (indicated mean effective pressure) meter, 135–136
    knock meter, 134–135
    PMax monitoring, 131–133, 133f, 134f, 135f
    properties, 124
  mechanical operation, 112
  purpose, 112
  sensor recognition, 131
  and SID (sensor identification)
    advantage, 130
    measurement procedure, 130
    surface acoustic wave (SAW) technology, 130, 130f
  and TEDS (Transducer Electronic Data Sheet)
    data security management, 127, 128f
    definition, 126
    installation, 129, 129f
    integration of, 127, 127f
    measurement procedure, 127
    mechanical operation, 128–129
    and microchips, 126–127, 126f
  tracking procedure, 130–131

Combustion measurement system, test environment
  components, 289–290
  configuration, 288
  conversion processes, 289f
  integration of, 288
  interfaces
    angle encoder to measurement system, 296–297
    charge amplifier to measurement system, 297, 298f
    components, 296, 298f
    data transfer and control, analogue, 300–301
    data transfer and control, digital, 299
    transducer to charge amplifier, 297
  systems
    high-end, 294–296, 294f
    low-end, 290–292, 290f
    mid-range, 292–294, 292f
Combustion noise
  definition, 242–244, 243f
  measurement procedures
    filters, 246–248, 246f, 247f
    human ear response, 247f
    signal processing and calculation, 244–246, 244f, 245f
  and quality considerations, 248
  significance of, 241–242
Combustion pressure curve
  definition, 23
  influence of crank angle resolution on, 24f, 24t
  measurement procedure, 23–24
  sampling frequency, 25
Crosshead beam engine indicator, 3f
Cylinder pressure measurements, combustion knock
  acquisition frequency, 225–226
  acquisition window, 226
  AVL histogram, 234–236, 235f
  AVL KI (knock index), 233, 233f
  AVL real time, 228–231, 230f
  AVL transient, 234
  calculation of knock overpressure, 226, 227f, 228

FEV CAS, 231–232, 233f
filtering, 224–225
frequency, 224
oscillation, 223f
rectification of peak curve, 228, 229f
signal processing, 224
third derivative, 236–237
weighting table, 232f
Cylinder pressure sensing, alternatives to pietzoelectric
  advantages, 92
  exhaust and inlet sensors
    acceleration effects, 108–109
    adaptors, 108f, 109f, 109–110, 111f
    installation, 108
    measurement procedure, 107–108, 107f
    purpose, 107
  optical sensors, as alternative to piezoelectric
    advantages, 94
    diagram, 94f
    installation, 94f
    measurement procedure, 93
    mechanical operation, 93
    requirements, 94
  pietzoresistant
    advantages, 92
    mechanical procedure, 93
  valve lift sensors
    laser vibrometry, 105–106, 106f
    measurement procedure, 106–107
    purpose, 105
    system, 106f
Cylinder pressure transducers
  construction and types
    cooled, 42–44, 43f, 44f, 45f
    design, 41, 42f
    and ground isolation, 41, 42f
    mechanical operation, 41
    uncooled, 44–47, 46f
  installation and adaptors
    considerations and positions, 62–66, 65f
    front sealing, 63f
    glow plug adaptors, 75–79, 77f, 78f
    installation of mounting bores, 66–69, 67f, 68f, 69f
    intervention, 62f
    intrusive mounting, 62
    nonintrusive mounting, 69–70
    spark plug adaptors, 70–75, 71f, 72f, 73f, 74f, 75f
  piezoelectric
    crystal cuts, 39, 40f
    direct, 36, 37f
    discovery, 36
    Gallium orthophosphate ($GAPO_4$), 39–40, 41
    materials used, 38–41
    mechanical procedure, 38
    purpose, 36
    quartz, 38–39
    reciprocal, 36–37, 37f
  selection and applications
    categories, 82–83
    considerations, 83–89, 83f, 85f, 89f
    requirements, 80–82, 81f
  transducer properties, piezoelectric
    design, 60–61
    environmental effects, 47, 48f
    specifications, 48–53, 49f
    thermodynamic, 54–60

Diesel engines. *See* Line pressure; Needle lift
Digital systems
  advantages, 15–16
  combustion measurement systems, 146
  data reduction, 17
  features, 15
  and future engine technologies, 17
  inputs and outputs, combustion measurement system hardware, 142
  measurement procedure, 16
  modern indicator system, 18f
  overview, 15f
  PC-based user interface, 17–18
  portable, 16–17, 16f
  real-time calculation ability, 17
  technology developments, 17–18
Drum-type engine indicator
  diagram produced by, 5f
  mechanical operation, 4, 4f

313

Eichelberg correlation, 191
Encoder output signal
    measurement procedure, 25–26
    mechanical operation, 25
    profile, typical, 26f
Encoder signal, processing
    accuracy of, 33
    and crank position sensor, 32f
    and crankshaft position information, 30
    limitations, 32
    measurement process, 32
    stages of testing, 31
    and toothed wheel, 30–31
    waveform display, 31f
Encoder types
    open, 28–29, 29f, 30f
    standard-closed, 27, 27f, 28f
Engine indicators
    analysis of diagrams
        measurement process, 12–13
        mechanical operation, 13
        planimeter, 12f, 13
    combustion measurement system, modern
        angle encoder, 20
        data acquisition system, 20
        dead weight tester, 20
        measurement procedure, 19
        signal conditioning amplifier, 19–20
        TDC (engine tope dead centre) sensor equipment, 20
        transducer, 19
        transducer conditioning system, 20
        workflow process, 19f
    crosshead beam, attached to, 3f
    digital systems
        advantages, 15–16
        data reduction, 17
        features, 15
        and future engine technologies, 17
        measurement procedure, 16
        modern indicator system, 18f
        overview, 15f
        PC-based user interface, 17–18
        portable, 16–17, 16f
        real-time calculation ability, 17
        technology developments, 17–18

drum-type indicator
    diagram produced by, 5f
    mechanical operation, 4, 4f
Farnsboro indicator
    cylinder pressure diagram, 11f
    pickup with disc valve arrangement, 10f
    recorder unit, 9f
    system, 8f
    and internal combustion engines, 6
Midgley optical engine indicator
    limitations, 8
    mechanical operation, 7
"moving tablet" indicator, 2, 3f
optical indicators
    diaphragm-type indicator, 6f
    limitations, 6–7
    mechanical operation, 6
oscilloscope recording
    components, 13
    measurement procedures, 14–15
    mechanical operation, 13–14
    system, 14f
    and transducer technology, 14–15
spark-trace indicator
    improvements in, 11
    "low-pressure" measurements, 11
    measurement procedure, 11
    mechanical operation, 8–10
Errors, sources of, in combustion measurement system
    amplifier, 259–260
    cabling, 257–258
    encoder, 258–259
    and measurement chain, 260–261, 261t
    measurement device, 260
    transducer, 257
Excel™, 150
Exhaust and inlet sensors
    acceleration effects, 108–109
    adaptors, 108f, 109f, 109–110, 111f
    installation, 108
    measurement procedure, 107–108, 107f
    purpose, 107

Farnsboro indicator
    cylinder pressure diagram, 11f

pickup with disc valve arrangement, 10f
recorder unit, 9f
system, 8f
Fuel systems. *See* Needle lift

Gallium orthophosphate (GAPO$_4$), 39–40, 41
 versus quartz, 41
General Motors, 8
Great Exhibition, 1862, 5

Hardware, combustion measurement system
 measurement data-flow sequence, 137–139
 operating requirements, 137
 system interfaces
  analogue inputs and outputs, 139–142
  angle encoder, 139
  digital inputs and outputs, 142
  operator interface, 139
Hiroyasu, Hiroyuki, 190

IEEE. *See* Institute of Electrical and Electronic Engineers (IEEE)
Ignition signals
 adjustment variables, 98
 angle measurement equipment, 99f
 definition, 97
 inductive clamp probe, 98, 98f
 mechanical operation, 99
Ignition timing amplifier, 121–122, 122f
Inductive angle encoder, 22–23
Installation and adaptors, piezoelectric. *See* Transducer installation and adaptors
Institute of Electrical and Electronic Engineers (IEEE), 126
Instrumentation, combustion measurement system, applications
 angle domain, 177
 angle encoders, 177
 external interfaces, 180
 hardware, 178–179
 pressure, 177–178
 signal conditioning, 178
Intelligent amplifiers
 advantages, 125–126
 analogue, 124
 for combustion noise, 133–134
 digital, 125, 125f
 IMEP (indicated mean effective pressure) meter, 135–136
 knock meter, 134–135
 PMax monitoring, 131–133, 133f, 134f, 135f
 properties, 124
 SID (sensor identification)
  advantage, 130
  measurement procedure, 130
  surface acoustic wave (SAW) technology, 130, 130f
 Transducer Electronic Data Sheet (TEDS)
  data security management, 127, 128f
  definition, 126
  installation, 129, 129f
  integration of, 127, 127f
  measurement procedure, 127
  mechanical operation, 128–129
  and microchips, 126–127, 126f
Interfaces, measurement system hardware
 analogue inputs and outputs, 139–142
 angle encoder, 139
 digital inputs and outputs, 142
 operator interface, 139
 typical system, 142–144, 143f
Interfaces, measurement system software
 controller area network (CAN), 161–163
 to engine electronics system, 159–160, 160f
 remote system, 157–159
Interfaces, PC-based, 17
Interfaces, and test environment, combustion measurement systems
 angle encoder to measurement system, 296–297
 charge amplifier to measurement system, 297, 298f
 components, 296, 298f
 data transfer and control, analogue, 300–301
 data transfer and control, digital, 299
 transducer to charge amplifier, 297
Interfaces, user, measurement system software
 basic function and requirements, 146
 data management, 149–151, 149f

Interfaces, user, measurement system software *(continued)*
  display of data, 148–149, 149f
  parameterisation, 147–148, 148f
Internal combustion engines, and impact on design of engine indicators, 6
Ion current sensors
  advantages, 95
  applications, 97
  in combustion chamber, 95f
  and glow plug design, 96f
  measurement procedure, 96–97
  mechanical operation, 95–96
  and pressure curve, 96f
  principle, basic, 95f
  purposes, 95

Jodon system, 197

The Kistler Company, 39
Knock, combustion
  and algorithm adaptations, 241
  causes, 219–220
  conditions producing, 221
  confusion about, 218
  cylinder pressure measurements
    acquisition frequency, 225–226
    acquisition window, 226
    AVL histogram, 234–236, 235f
    AVL KI (knock index), 233, 233f
    AVL real time, 228–231, 230f
    AVL transient, 234
    calculation of knock overpressure, 226, 227f, 228
    FEV CAS, 231–232, 233f
    filtering, 224–225
    frequency, 224
    oscillation, 223f
    rectification of peak curve, 228, 229f
    signal processing, 224
    third derivative, 236–237
    weighting table, 232f
  definition, 218–219, 219f
  measurement techniques
    instrumented spark plug, 222–223
    ionization current, 222
    optical access, 222
    structure-bourne knock sensors, 221–222
  quality considerations
    measurement range and resolution, 239
    measurement system setup, 239–240
    transducer position, type, and properties, 237–238, 238f, 239f
  significance of, 220–221

Line pressure
  challenges in the measurement of, 101
  dynamic effects, 99
  installation, 100, 100f
  measurement procedure, 100
  mechanical operation, 100
Low pressure-gas exchange process
  and low pressure transducers, 214–216, 215f
  measurement, analysis, 213–214, 214f
  measurement, setup, 212–213, 213f
  measurement, task and goal, 212–212
  mechanical operation, 209–210

Maintenance, combustion measurement system
  encoder, 276–278, 277t
  hardware and system, 275–276
  transducers
    calibration, 280–284, 281f, 282f, 283f
    installation and handling, 278–279
    maintenance and repair, 279–280
    quality considerations, 278
Matlab™, 150
Measurement procedures, combustion noise
  filters, 246–248, 246f, 247f
  human ear response, 247f
  signal processing and calculation, 244–246, 244f, 245f
Measurement system software
  calculations and results
    derived results, importance of, 163
    future developments, 169–170
    real-time results, 164–165
    user-defined results, 165–169, 167f, 168f

combustion measurement system
  best practices, 274–275
  measured data requirements, 269–272
  plausibility checks and data validation, 272–273, 273f, 274f
  reliability, 269
interfaces
  controller area network (CAN), 161–163
  to engine electronics system, 159–160, 160f
  remote system, 157–159
operating modes
  advanced thermodynamic analysis, 156
  cold start, 153
  event mode, 154
  fuel injection analysis, 156
  knock measurement, 153
  optical and flame analysis, 156–157
  pulse frequency analysis, 157
  standard measurement options, 151–152, 152f
  time-based, 154–155
  torsion and rotation analysis, 155, 156f
postprocessing and data management
  basic requirements for data format and export, 170–172
  main purpose, 170
  requirements for engine and system parameters, 172–173
  typical environment, 173–174
user interface
  basic function and requirements, 146
  data management, 149–151, 149f
  display of data, 148–149, 149f
  parameterisation, 147–148, 148f
Microsoft Visual Basic™, 150
Microsoft Windows™, 146
Midgley optical engine indicator, 7f
  limitations, 8
  mechanical procedure, 7
Midgley, Thomas, Jr., 7
MOSFET (metal oxide semiconductor field effect transistor), 38
"Moving tablet" indicator measurement procedure, 2

Needle lift
  amplification, 103, 105
  calibration, 105
  definition, 101
  Hall effect sensing element, 102, 103f
  inductive method, 101
  installation, 103
  instrumented injectors, 103f
  measurement procedure, 104
  production equipment, 101–102, 102f
  and signal output, 104–105, 104f
  technologies for, 102
Noise, combustion
  definition, 242–244, 243f
  measurement procedures
    filters, 246–248, 246f, 247f
    human ear response, 247f
    signal processing and calculation, 244–246, 244f, 245f
  and quality considerations, 248
  significance of, 241–242

Operating modes, measurement system software
  advanced thermodynamic analysis, 156
  cold start, 153
  event mode, 154
  fuel injection analysis, 156
  knock measurement, 153
  optical and flame analysis, 156–157
  pulse frequency analysis, 157
  standard measurement options, 151–152, 152f
  time-based, 154–155
  torsion and rotation analysis, 155, 156f
Optical angle encoder, 22
Optical indicators
  diaphragm-type indicator, 6f
  limitations, 6–7
  mechanical operation, 6
Optical sensors, as alternative to piezoelectric
  advantages, 94
  diagram, 94f
  installation, 94f
  measurement procedure, 93

Optical sensors, as alternative to *(continued)*
   mechanical operation, 93
   requirements, 94
Oscilloscope recording
   components, 13
   measurement procedures, 14–15
   mechanical procedure, 13–14
   system, 14f
   and transducer technology, 14–15

PC-based user interface, 17
Piezoelectric transducers
   crystal cuts, 39, 40f
   direct, 36, 37f
   discovery, 36
   Gallium orthophosphate ($GAPO_4$), 39–40, 41
   materials used, 38–41
   mechanical operation, 38
   purpose, 36
   quartz, 38–39
   reciprocal, 36–37, 37f
Piezoresistant sensors
   advantages, 92
   mechanical procedure, 93
Planimeter, 12f
Porter-Allen engine, 5
Postprocessing and data management, measurement system software
   basic requirements for data format and export, 170–172
   main purpose, 170
   requirements for engine and system parameters, 172–173
   typical environment, 173–174
Prechecks, combustion measurement system
   amplifiers, 263
   encoder, 263–264
   measurement procedure, 264
   measurement setup, 262
   transducer and cabling, 262–263

Quartz, 39
   versus Gallium orthophosphate ($GAPO_4$), 39–40, 41

Rassweiler, G.M., 201–202
Richards indicator, 5, 5f

SID (sensor identification)
   advantage, 130
   measurement procedure, 130
   surface acoustic wave (SAW) technology, 130, 130f
Signal conditioning systems for combustion measurement
   and amplifier systems, 112–113, 113f
   analogue signals
      mechanical operation, 118–119
      types of signals, 119, 120f, 121
   carrier-frequency amplifier, 122–124, 123f, 124f
   charge amplifier
      basic function and operation, 113
      cabling and interfaces to, 117
      design, 113, 115f
      drift and drift compensation, 116–117, 116f, 117f
      measurement procedure, 113–114
      piezoelectric sensor, 113
      and time constant property, 115
   design, 112, 113f, 114f
   harmonisation of elements, 131
   ignition timing amplifier, 121–122, 122f
   information provided, 130
   intelligent amplifiers
      advantages, 125–126
      analogue, 124
      for combustion noise, 133–134
      digital, 125, 125f
      IMEP (indicated mean effective pressure) meter, 135–136
      knock meter, 134–135
      PMax monitoring, 131–133, 133f, 134f, 135f
      properties, 124
   mechanical operation, 112
   purpose, 112
   sensor recognition, 131
   and SID (sensor identification)
      advantage, 130
      measurement procedure, 130

surface acoustic wave (SAW)
        technology, 130, 130f
    and TEDS (Transducer Electronic Data
        Sheet)
        data security management, 127, 128f
        definition, 126
        installation, 129, 129f
        integration of, 127, 127f
        measurement procedure, 127
        mechanical operation, 128–129
        and microchips, 126–127, 126f
    tracking procedure, 130–131
Software and data handling
    best practices, 274–275
    measured data requirements, 269–272
    plausibility checks and data validation,
        272–273, 273f, 274f
    reliability, 269
Software, measurement system
    derived results, importance of, 163
    future developments, 169–170
    real-time results, 164–165
    user-defined results, 165–169, 167f, 168f
Spark-trace indicator. *See also* Farnsboro
        indicator
    improvements in, 11
    "low-pressure" measurements, 11
    measurement procedure, 11
    mechanical operation, 8–10
Steam engine pioneers, 2
Stirling engines, 197
System interfaces, combustion measurement
        system
    analogue inputs and outputs, 139–142
    angle encoder, 139
    digital inputs and outputs, 142
    operator interface, 139
    typical system, 142–144, 143f

TDC (top dead centre) combustion
        measurement system application
    capacitive probe, 191–196. *See also*
        Capacitive probe
    determination methods, 197–200, 199f
    interval method, 190f
    introduction, 185–186, 186f
    microwave, 197
    motored curve method, 191
    pressure curve determination, 188–191
    static determination, 186–187, 187f
    thermodynamic loss angle, 187f, 188f
TEDS. *See* Transducer Electronic Data Sheet
        (TEDS)
Test environment, combustion measurement
        system
    components, 289–290
    configuration, 288
    conversion processes, 289f
    integration of, 288
    interfaces
        angle encoder to measurement system,
            296–297
        charge amplifier to measurement
            system, 297, 298f
        components, 296, 298f
        data transfer and control, analogue,
            300–301
        data transfer and control, digital,
            299
        transducer to charge amplifier, 297
    systems
        high-end, 294–296, 294f
        low-end, 290–292, 290f
        mid-range, 292–294, 292f
Thermodynamic analysis, combustion
        measurement system applications
    calculations, offline, 206–207, 207f
    definition, 200–201
    error-producing effects, 208–209
    heat release algorithm, 203–206, 204f
    measurement procedure, 203–206
    mechanical operation, 202–203
Transducer calibration, 280–284, 281f, 282f,
        283f
Transducer combustion measurement
        applications
    ignition signals
        adjustment variables, 98
        angle measurement equipment, 99f
        definition, 97
        inductive clamp probe, 98, 98f
        mechanical operation, 99

Transducer combustion measurement
    applications *(continued)*
  line pressure
    challenges in the measurement of, 101
    dynamic effects, 99
    installation, 100, 100f
    measurement procedure, 100
    mechanical operation, 100
  needle lift
    amplification, 103, 105
    calibration, 105
    definition, 101
    Hall effect sensing element, 102, 103f
    inductive method, 101
    installation, 103
    instrumented injectors, 103f
    measurement procedure, 104
    production equipment, 101–102, 102f
    and signal output, 104–105, 104f
    technologies for, 102
Transducer construction and types, piezoelectric
  cooled, 42–44, 43f, 44f, 45f
  design, 41, 42f
  and ground isolation, 41, 42f
  mechanical operation, 41
  uncooled, 44–47, 46f
Transducer, cylinder pressure
  construction and types
    cooled, 42–44, 43f, 44f, 45f
    design, 41, 42f
    and ground isolation, 41, 42f
    mechanical operation, 41
    uncooled, 44–47, 46f
  installation and adaptors
    considerations and positions, 62–66, 65f
    front sealing, 63f
    glow plug adaptors, 75–79, 77f, 78f
    installation of mounting bores, 66–69, 67f, 68f, 69f
    intervention, 62f
    intrusive mounting, 62
    nonintrusive mounting, 69–70
    spark plug adaptors, 70–75, 71f, 72f, 73f, 74f, 75f

  piezoelectric
    crystal cuts, 39, 40f
    direct, 36, 37f
    discovery, 36
    Gallium orthophosphate ($GAPO_4$), 39–40, 41
    materials used, 38–41
    mechanical procedure, 38
    purpose, 36
    quartz, 38–39
    reciprocal, 36–37, 37f
  selection and applications
    categories, 82–83
    considerations, 83–89, 83f, 85f, 89f
    requirements, 80–82, 81f
  transducer properties, piezoelectric
    design, 60–61
    environmental effects, 47, 48f
    specifications, 48–53, 49f
    thermodynamic, 54–60
Transducer Electronic Data Sheet (TEDS)
  data security management, 127, 128f
  definition, 126
  installation, 129, 129f
  integration of, 127, 127f
  measurement procedure, 127
  mechanical operation, 128–129
  and microchips, 126–127, 126f
Transducer installation and adaptors, 62–78
  considerations and positions, 62–66, 65f
  front sealing, 63f
  glow plug adaptors, 75–79, 77f, 78f
  installation of mounting bores, 66–69, 67f, 68f, 69f
  intervention, 62f
  intrusive mounting, 62
  nonintrusive mounting, 69–70
  spark plug adaptors, 70–75, 71f, 72f, 73f, 74f, 75f
Transducer properties, piezoelectric
  design, 60–61
  environmental effects, 47, 48f
  specifications, 48–53, 49f
  thermodynamic, 54–60, 55f, 56f, 58f, 59f, 61

Transducer selection and applications, 79–89
    categories, 82–83
    considerations, 83–89, 83f, 85f, 89f
    requirements, 80–82, 81f
Transducer technology
    inductive, 14
    piezoelectric, 14
    strain gauge, 15
    variable capacitance, 14–15

User interface, measurement system software
    basic function and requirements, 146
    data management, 149–151, 149f
    display of data, 148–149, 149f
    parameterisation, 147–148, 148f

Valve lift sensors
    laser vibrometry, 105–106, 106f
    measurement procedure, 106–107
    purpose, 105
    system, 106f

Watt, James, 2
Withrow, L., 201–202
Wolff Company, 102
Woschni correlation, 191

Zero-level correction, combustion measurement system applications
    detection methods, 184f
    fixed point and measured value, 180–181, 181f
    fixed point and reference value, 179–180, 180f
    introduction, 179
    post-processing, 183–184
    thermodynamic, 181–183, 182f

ABOUT THE AUTHOR

David R. Rogers started with a technical career in the retail motor industry. He then moved into technical positions in automotive research and development and following this, a position in technical sales and business management for automotive powertrain instrumentation and test systems allowed him to develop a detailed, practical knowledge base on combustion pressure measurement, from a user perspective and application viewpoint. He gained much of this knowledge through hands-on interaction with users of the equipment in real-life applications.

Professionally, he is a member of the Institution of Mechanical Engineers and the Society of Automotive Engineers. He is also currently a member of the board of the Institution of Mechanical Engineers, Automobile Division.